Math Mammoth
Introduction to Fractions

By Maria Miller

Copyright 2025 Taina Maria Miller
ISBN 978-1-954358-58-4

2025 EDITION

All rights reserved. No part of this book may be reproduced or transmitted in any form or by any means, electronic or mechanical, or by any information storage and retrieval system, without permission in writing from the author.

Copying permission: For having purchased this book, the copyright owner grants to the teacher-purchaser a limited permission to reproduce this material for use with his or her students. In other words, the teacher-purchaser MAY make copies of the pages, or an electronic copy of the PDF file, and provide them at no cost to the students he or she is actually teaching, but not to students of other teachers. This permission also extends to the spouse of the purchaser, for the purpose of providing copies for the children in the same family. Sharing the file with anyone else, whether via the Internet or other media, is strictly prohibited.

No permission is granted for resale of the material.

The copyright holder also grants permission to the purchaser to make electronic copies of the material for back-up purposes.

If you have other needs, such as licensing for a school or tutoring center, please contact the author at
https://www.MathMammoth.com/contact

Contents

Introduction	5
Halves and Quarters	9
Some Fractions	13
Understanding Fractions 1	16
Understanding Fractions 2	20
Fractions on a Number Line 1	24
Fractions on a Number Line 2	27
One Whole and Its Fractional Parts	31
Mixed Numbers	34
Mixed Numbers and Fractions	38
Comparing Fractions 1	41
Comparing Fractions 2	43
Comparing Fractions 3	46
Comparing Fractions 4	49
Comparing Fractions 5	51
Equivalent Fractions 1	55
Equivalent Fractions 2	57
Equivalent Fractions 3	59
Equivalent Fractions 4	61
Adding Fractions	66
Adding Mixed Numbers	68
Subtracting Fractions and Mixed Numbers	71
Multiplying Fractions by Whole Numbers	75
Practicing with Fractions	78
Fractions Review	80
Review	83
Answers	85
More from Math Mammoth	119

Introduction

Math Mammoth Introduction to Fractions contains lessons for fraction arithmetic for grades 1-4. The bulk of the material here is for third and fourth grades, and only a few lessons are intended for grades 1-2.

The topics covered are on a simple level, illustrated with visual models, and with small denominators. The presentation avoids spelling out specific rules for manipulating fractions, but instead relies on the usage of pictures on a very concrete level. Children easily confuse the various rules for fraction arithmetic, because there are so many. There is a place for the rules, as *shortcuts* for ideas that are already understood, but we do not start with them. The goal is to let the big ideas sink in conceptually first, followed by some shortcuts.

The topics covered are

- one-half and one-fourth
- concept of a fraction
- concept of a mixed number
- comparing fractions
- equivalent fractions
- adding and subtracting like fractions
- adding and subtracting mixed numbers with like fractional parts
- adding one fraction that has tenths and another that has hundredths (such as 3/10 + 7/100)
- multiplying a fraction by a whole number

The book does not cover division or multiplication of fractions, nor adding unlike fractions, which are topics for fifth and sixth grades.

The lessons are organized by topic, not by increasing difficulty. For reference, first grade students study only the concept of one-half and one-fourth. In second grade, they study the concept of a fraction and optionally the easiest (first) lesson on comparing fractions. In third grade, students study the concept of a fraction, fractions on a number line, comparing fractions, and equivalent fractions. Then in fourth, they study mixed numbers, comparing fractions, equivalent fractions, adding and subtracting fractions and mixed numbers, and multiplying fractions by whole numbers.

The answers are appended.

I wish you success in teaching math!

Maria Miller, the author

Games and Activities

Make It Less! (Game for comparing fractions)

You need: A deck or two of number cards from 1-10. (A standard deck of cards will work if you remove the face cards.) One person to be the judge, checking the players' math work.

Game play: Deal four cards to each player. Place the rest of the cards as a deck face down on the table. In the beginning of each round, two cards are drawn from the deck to form the reference fraction for that round. This will always be a proper fraction or a fraction equal to 1. For example, if 3 and 9 are drawn from the deck, the reference fraction is 3/9 (not 9/3). If 5 and 5 are drawn, the reference fraction is 5/5 (which makes for a very easy round for the players!). For each round there will be a new reference fraction. After the round is over, mix the cards from the reference fraction back in with the other cards in the deck.

Then, each player tries to form as many fractions as they can that are *less than* the reference fraction, using the cards in their hand (one card for the numerator and one for the denominator).

If the judge agrees, the player gets to put those two cards in their personal pile. (If the judge does not agree, nothing happens — the player will neither draw more cards nor get to put any in their personal pile.)

If the player cannot form any such fraction, they draw one card from the deck.

At the end of the round, players who have less than four cards in their hand take enough cards from the deck to again have four cards in their hand.

Some examples:

Say the reference fraction is 2/5 and you have 2, 6, 8, and 1. You could make the fractions 1/6 and 2/8, and get to put all four cards into your personal pile, since both 1/6 and 2/8 are less than 2/5.

Or, say the reference fraction is 1/6 and you have 8, 7, 8, and 4. You cannot form any fraction that is less than 1/6, thus you draw one card from the deck and your turn is over.

Or, say the reference fraction is 7/9 and you have 4, 5, 9, 9, and 8. You can form 4/9 and 5/8, leaving just one 9 in your hand. You get to put 4, 9, 5, and 8 in your personal pile and you will then draw three new cards from the deck.

The game continues until all the cards from the deck are used. The player with the most cards in their personal pile wins.

Games and Activities at Math Mammoth Practice Zone

Fraction Matcher
In this activity, you match visual models of fractions (or mixed numbers) with other visual models, or with the fractions written as numbers. For third grade, choose to work with fractions (not mixed numbers).
https://www.mathmammoth.com/practice/fraction-matcher

Fractions on a number line
Practice marking fractions on a number line with this interactive exercise. For third grade, choose to use both proper and improper fractions.
https://www.mathmammoth.com/practice/fraction-number-line

Equivalent Fractions with Visual Models
Explore equivalent fractions visually with this interactive online activity and a matching game.
https://www.mathmammoth.com/practice/fractions-equality

Order Fractions on a Number Line
In this activity, you will place three given fractions on a number line. The fractions have the same denominator.
https://www.mathmammoth.com/practice/order-fractions-number-line#questions=5&number=propFrac,impropFrac&sameDen=1

Mixed Numbers with Visual Models
Explore mixed numbers and improper fractions with the help of visual models and the number line. You can also play a game where you make mixed numbers from visual models.
https://www.mathmammoth.com/practice/mixed-numbers

Equivalent Fractions Hidden Picture Matching Game
Practice simplifying fractions to find equivalent fractions while also uncovering a hidden picture in this fun matching game!
https://www.mathmammoth.com/practice/equivalent-fractions

Compare Fractions
Practice comparing fractions. In the link below, you will compare fractions that have the same denominator, same numerator, or one of the fractions is 1/2 or equal to 1.
https://www.mathmammoth.com/practice/comparing-fractions#q=10&dens=2,3,4,5,6,8&pie=0&improper=0&types=0,1,2,3

Add and Subtract Fractions
Practice addition and subtraction of fractions.
https://www.mathmammoth.com/practice/add-fractions#like=1&proper=1&simplified=0¬-lowest=1&denominators=2,3,4,5,6,8

Multiply Fractions
On this page, you can practice multiplying fractions by whole numbers.
https://www.mathmammoth.com/practice/multiply-fractions#proper=0¬-lowest=1&level=easy&multiply=fbwn

Helpful Resources on the Internet

We have compiled a list of Internet resources that match the topics in this book. This list of links includes web pages that offer:

- **online practice** for concepts;
- online **games**, or occasionally, printable games;
- **animations** and interactive **illustrations** of math concepts;
- **articles** that teach a math concept.

We heartily recommend you take a look at the list. Many of our customers love using these resources to supplement the bookwork. You can use the resources as you see fit for extra practice, to illustrate a concept better, and even just for some fun. Enjoy!

https://l.mathmammoth.com/blue/introductiontofractions

Halves and Quarters

This square is divided into two parts that are the same. The parts are *halves*. Each part is one-half.		This circle is divided into four parts that are the same. The parts are called *fourths* or *quarters*. Each part is one fourth or one quarter.	
Here, one half of the square is colored. The other half is white.		Here, three fourths of the circle are colored. One-fourth of it is white.	

1. Divide these shapes into halves by drawing a straight line from dot to dot. Then color as you are asked to.

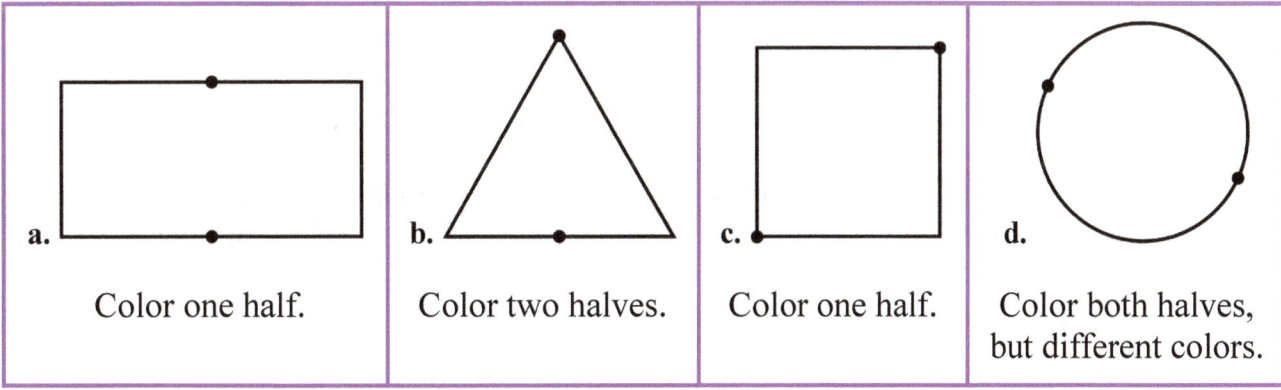

a. Color one half. b. Color two halves. c. Color one half. d. Color both halves, but different colors.

2. Divide these shapes into fourths by drawing two straight lines from dot to dot. Then color as you are asked to.

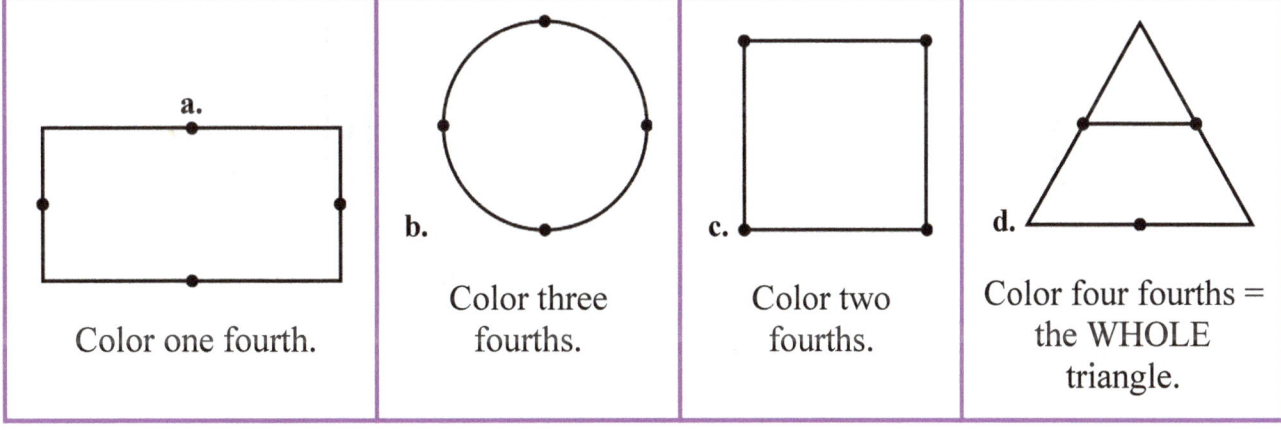

a. Color one fourth. b. Color three fourths. c. Color two fourths. d. Color four fourths = the WHOLE triangle.

3. Color. Then compare.

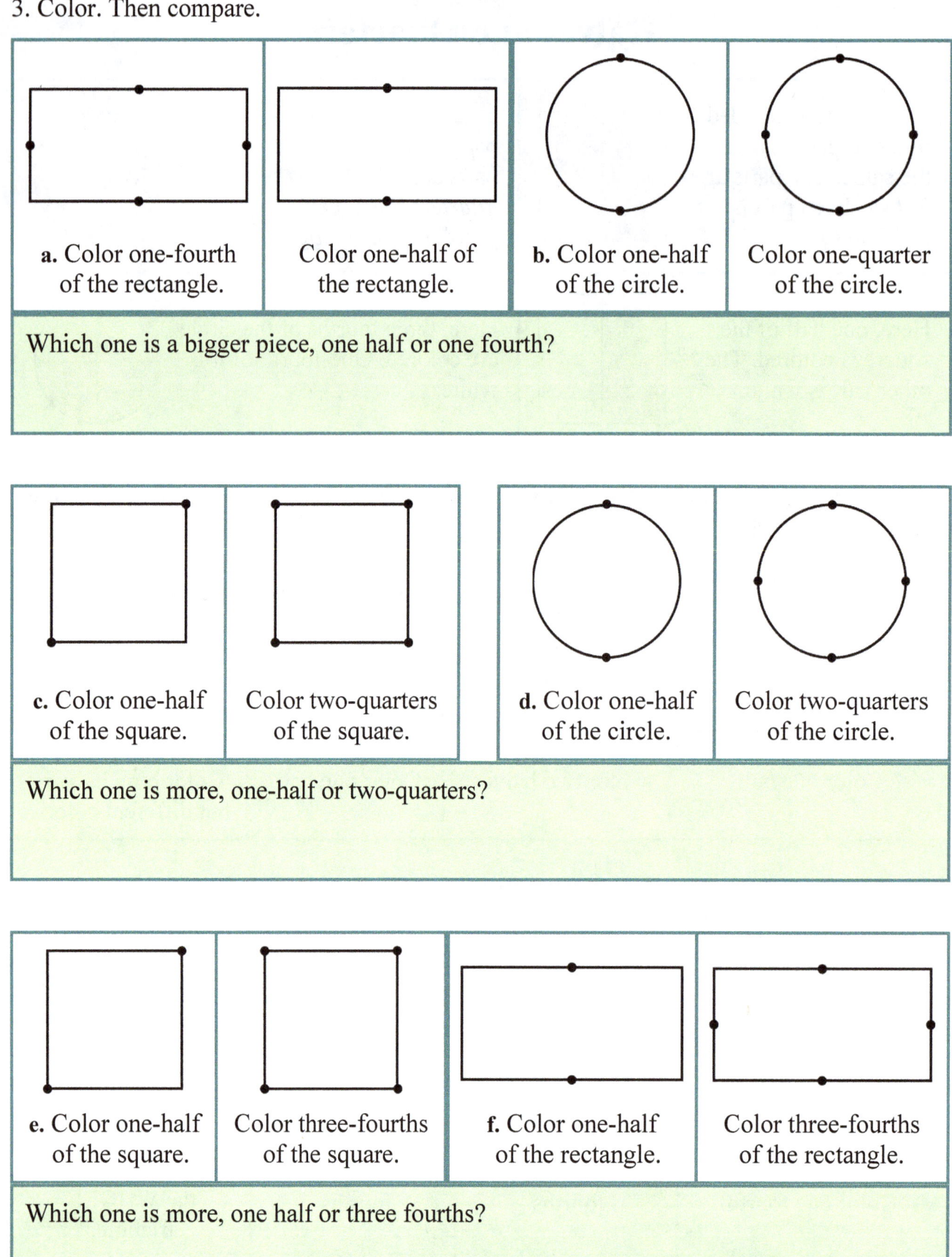

This square is divided into three parts that are the same. The parts are *thirds*. Each part is <u>one third</u>.

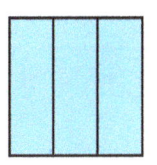

4. Color.

a.
Color one third.

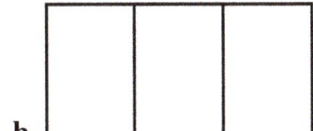
b.
Color two thirds.

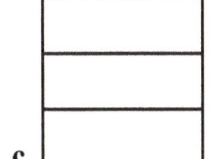
c.
Color one third.

d.
Color three thirds.

5. Color. Then compare.

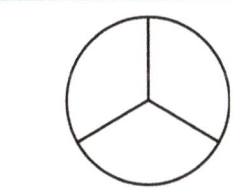
Color two thirds.   Color one half.

a. Which is more, two thirds or one half?

Color three fourths.   Color two thirds.

b. Which is more, three fourths or two thirds?

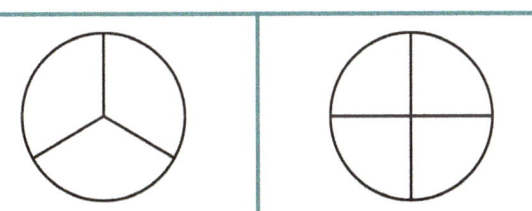

c. Which is more, two thirds or two quarters?

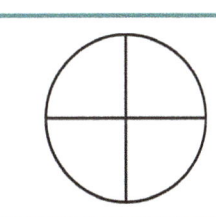

d. Which is more, two fourths or one half?

6. Color ONE piece in each pie. Then compare. Think of eating pie pieces!

a. Which is more, one half or one third?

b. Which is more, one fourth or one third?

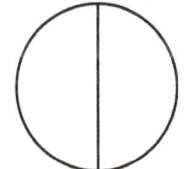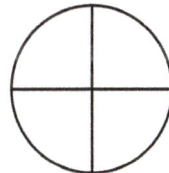

7. Color the whole pie. Then tell or write how many pieces it is, and what kind of pieces.

a. The whole pie is __3__ __thirds__ .

b. The whole pie is _____ _____ .

c. The whole pie is _____ _____ .

8. Tell or write how many pieces, what kind of pieces, and of what shape are colored. Look at the example.

a. __1__ __fourth__ of the __oval__ is colored.

b. _____ _____ of the __hexagon__ is colored.

c. _____ _____ of the __trapezoid__ is colored.

d. _____ _____ of the _____ are colored.

e. _____ _____ of the _____ are colored.

f. _____ _____ of the _____ are colored.

Some Fractions

We will now divide shapes into EQUAL parts = parts that are the same size.
When we divide something into TWO equal parts, the parts are called *halves*.
When we divide something into THREE equal parts, the parts are called *thirds*.
When we divide something into FOUR equal parts, the parts are *fourths* or *quarters*.

Here, one-half of the square is colored. We write $\frac{1}{2}$ or 1/2.

Here, two-halves of the square are colored. We write $\frac{2}{2}$ or 2/2. This is the same as 1 (one whole).

Here, one-third of the square is colored. We write $\frac{1}{3}$ or 1/3.

Now, four-quarters of the circle are colored. We write $\frac{4}{4}$ or 4/4. This is the same as 1 (one whole).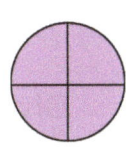

In a *fraction,* we use two numbers, one on the top and one on the bottom.

One-fourth of the pie is colored.

how many parts colored → $\frac{1}{4}$
how many equal parts →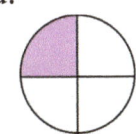

Two-thirds of the square is colored.

how many parts colored → $\frac{2}{3}$
how many equal parts →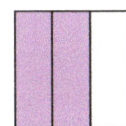

1. Divide these shapes. Then color as you are asked to.

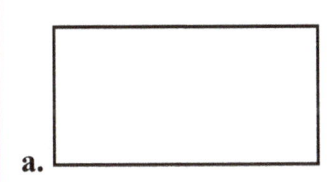
a.
Divide this into halves. Color $\frac{1}{2}$

b.
Divide this into thirds. Color $\frac{1}{3}$

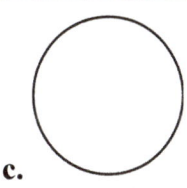
c.
Divide this into halves. Color $\frac{2}{2}$

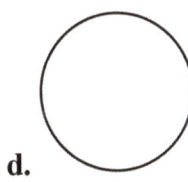
d.
Divide this into fourths. Color $\frac{2}{4}$

e.
Divide this into quarters. Color $\frac{4}{4}$

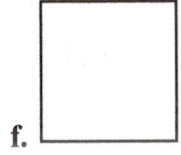
f.
Divide this into thirds. Color $\frac{2}{3}$

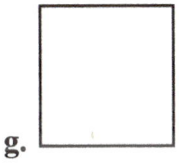
g.
Divide this into fourths. Color $\frac{1}{4}$

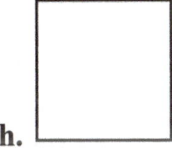
h.
Divide this into halves. Color $\frac{2}{2}$

Robert divided this square into fourths, and then colored $\frac{1}{4}$ of it.

<u>Notice</u>: the whole rectangle has 16 **little squares** inside it.

The fourth that Robert colored has 4 **little squares** inside it.

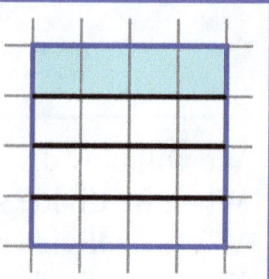

2. Complete.

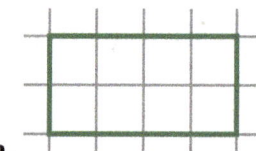

a.

Divide this into halves. Color $\frac{1}{2}$

_____ little squares in one half.

_____ little squares in the whole rectangle

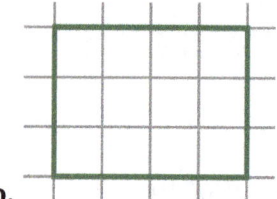

b.

Divide this into halves. Color $\frac{1}{2}$

_____ little squares in one half.

_____ little squares in the whole rectangle

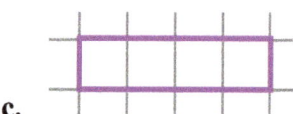

c.

Divide this into fourths. Color $\frac{1}{4}$

_____ little squares in one fourth.

_____ little squares in the whole rectangle

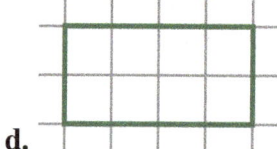

d.

Divide this into fourths. Color $\frac{1}{4}$

_____ little squares in one fourth.

_____ little squares in the whole rectangle

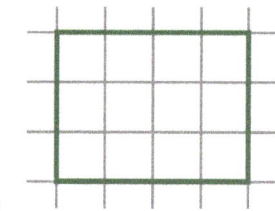

e.

Divide this into fourths. Color $\frac{3}{4}$

_____ little squares in three fourths.

_____ little squares in the whole rectangle

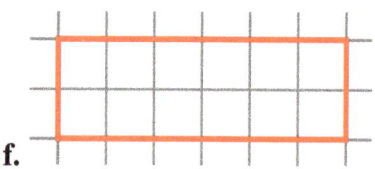

f.

Divide this into thirds. Color $\frac{2}{3}$

_____ little squares in two thirds.

_____ little squares in the whole rectangle

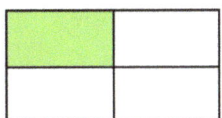 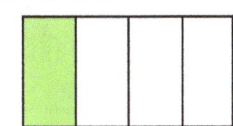

Margie divided a rectangle into quarters one way, and then colored $\frac{1}{4}$.	Jessie divided a rectangle into quarters another way, and then colored $\frac{1}{4}$.

Which one is MORE? Well, they are both one-fourth! So, they are <u>equal</u>.

THINK: If you had a chocolate bar cut into quarters Margie's way or Jessie's way, and you got 1/4, either way you would get to *eat* the same amount.

3. The dots show you how to divide the shape. Divide it, then color.

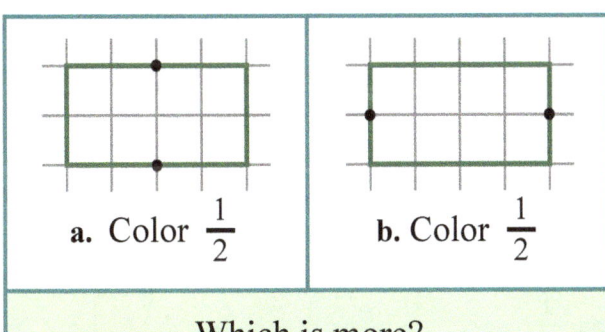 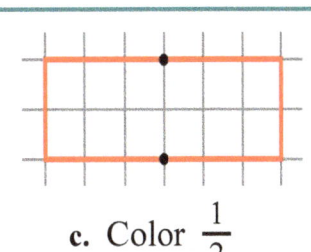 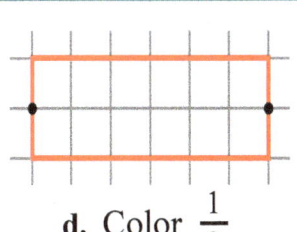

a. Color $\frac{1}{2}$ **b.** Color $\frac{1}{2}$ **c.** Color $\frac{1}{2}$ **d.** Color $\frac{1}{2}$

Which is more?   Which is more?

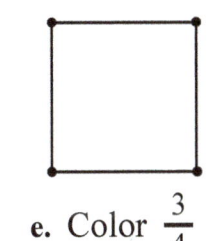 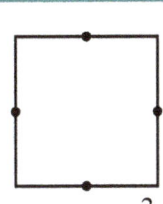 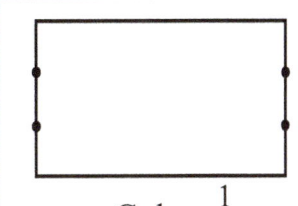 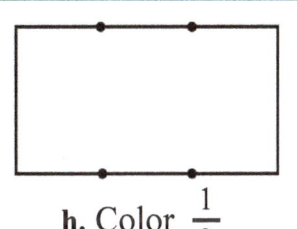

e. Color $\frac{3}{4}$ **f.** Color $\frac{3}{4}$ **g.** Color $\frac{1}{3}$ **h.** Color $\frac{1}{3}$

Which is more?   Which is more?

4. Tell what fraction is colored.

   **a.**   **b.**   **c.**   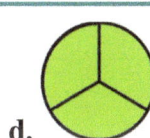 **d.**

Understanding Fractions 1

Example: Three friends share a pizza. Here you see how the pizza is divided into three parts. Would you say that each person will get a fair share?

Why or why not?

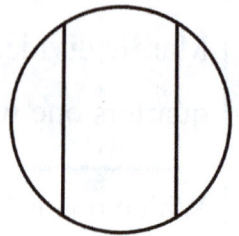

   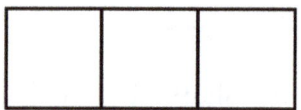   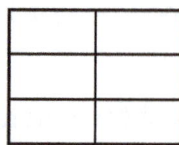

Here, one whole is divided into two <u>equal</u> parts. Each part is one-half of the whole.

Here, one whole is divided into three <u>equal</u> parts. Each part is one-third of the whole.

Here, one whole is divided into six <u>equal</u> parts. Each part is one-sixth of the whole.

One whole is divided into four equal parts. One part is colored.

That one part is 1 fourth of the whole, and is written as $\frac{1}{4}$.

One whole is divided into eight equal parts. One part is colored.

That one part is 1 eighth of the whole, and is written as $\frac{1}{8}$.

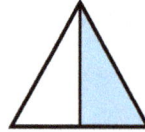 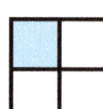  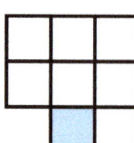 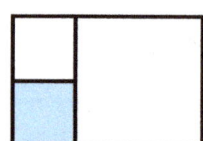

1 half   1 third   1 sixth   1 eighth   Here, the whole is not divided into equal parts, so we cannot easily tell what fraction this is.

$\frac{1}{2}$ $\frac{1}{3}$ $\frac{1}{6}$ $\frac{1}{8}$

The fractions $\frac{1}{2}, \frac{1}{3}, \frac{1}{4}, \frac{1}{6}, \frac{1}{8}$ and so on are called **unit fractions**.

A unit fraction signifies ONE part of a whole, when the whole is divided into *equal* parts.

1. In each picture, one part is shaded. **If** the one whole is divided into *equal* parts, write the fraction that is formed, and otherwise not.

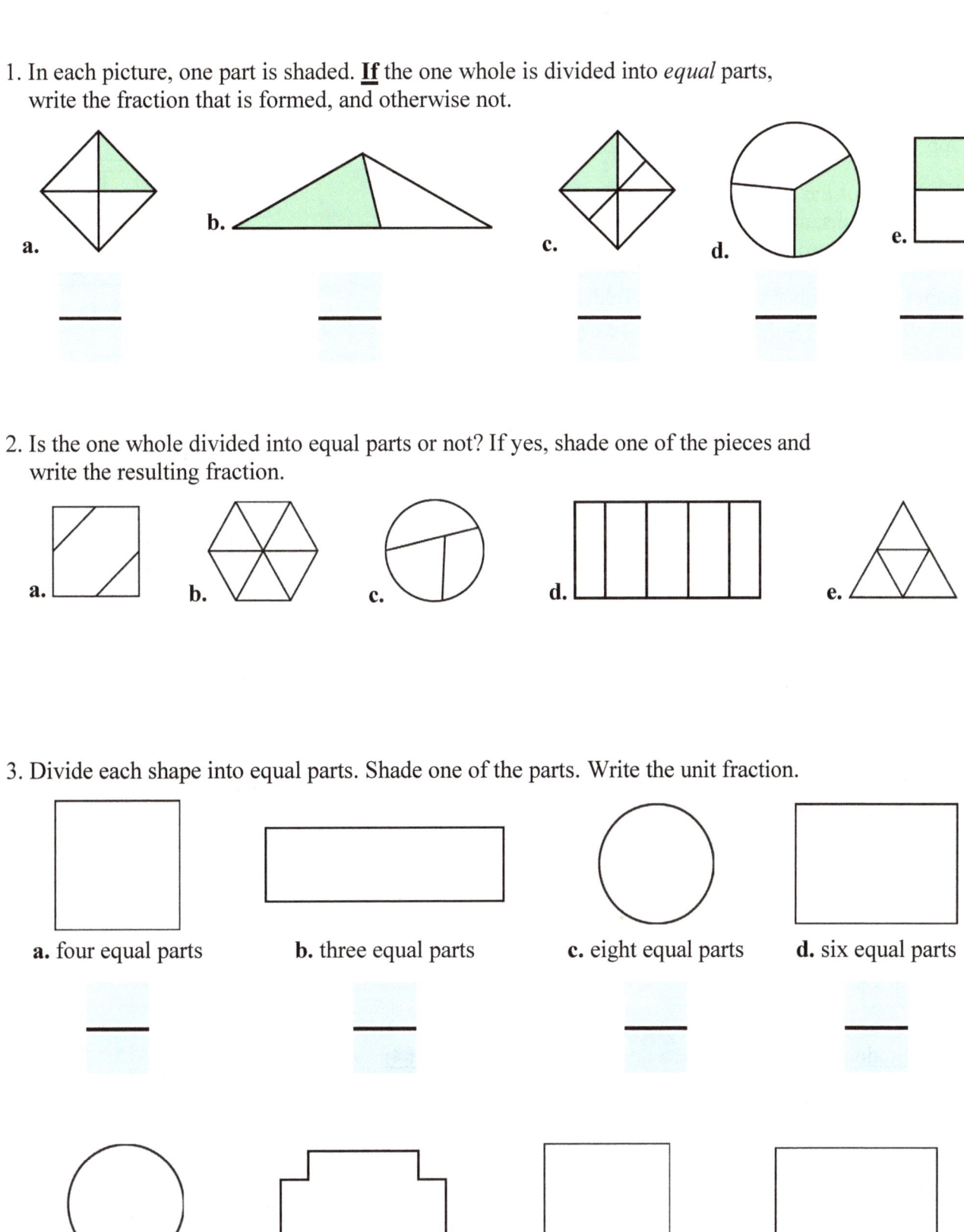

2. Is the one whole divided into equal parts or not? If yes, shade one of the pieces and write the resulting fraction.

3. Divide each shape into equal parts. Shade one of the parts. Write the unit fraction.

a. four equal parts   b. three equal parts   c. eight equal parts   d. six equal parts

e. four equal parts   f. two equal parts   g. eight equal parts   h. three equal parts

4. **Activity: fraction strips**

   You will need: Five identical strips of paper, approximately 6 inches long and 1 inch tall.

   Fold three of the strips of paper so that you will get halves, fourths, and eighths. Fold the other two so that you will get thirds and sixths. See the illustration.

   Label the individual unit fractions. In the illustration, one of the thirds is marked. Use pretty colors.

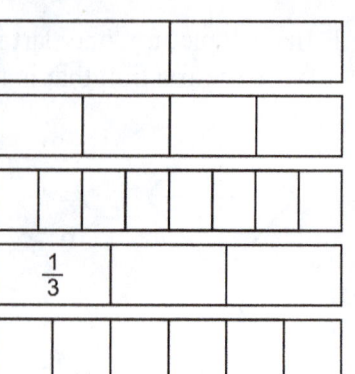

5. Mandy drew a shape, divided it into equal parts, and colored $\frac{1}{6}$ of it.

   Which ones of these shapes below could be the one she drew?

   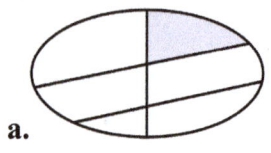 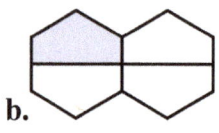 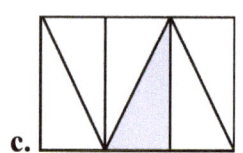
   a.   b.   c.

    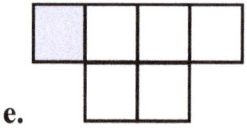 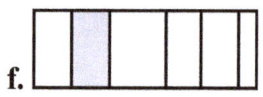
   d.   e.   f.

6. Students wrote what fraction of the shape is shaded. Which students made an error? Explain why they are in error.

Student 1: 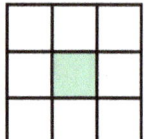 $\frac{1}{8}$	Student 2: 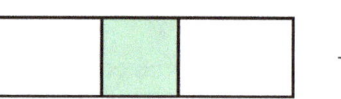 $\frac{1}{3}$
Student 3: 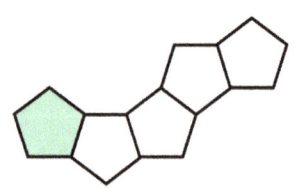 $\frac{1}{5}$	Student 4: 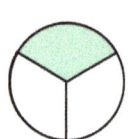 $\frac{1}{2}$
Student 5: 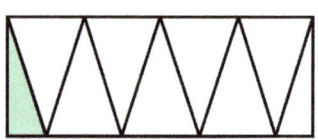 $\frac{1}{9}$	Student 6: 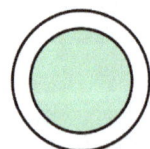 $\frac{1}{2}$

Use questions 7 and 8 for extra practice.

7. IF the shape is divided into equal parts, shade one of the parts. Then write the fraction.

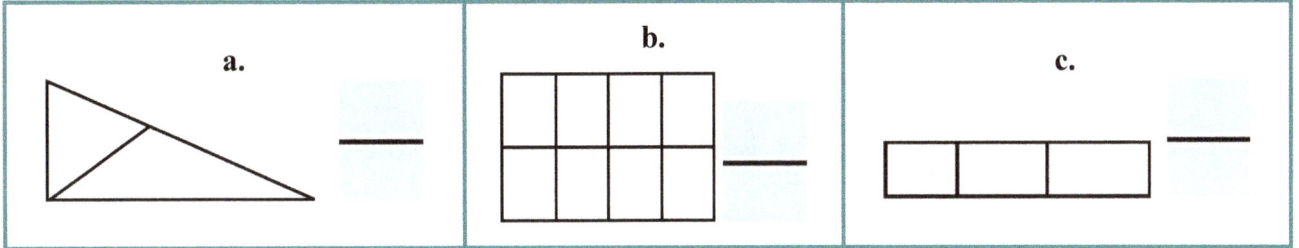

8. Divide each shape into equal parts. Shade one of the parts. Write the unit fraction.

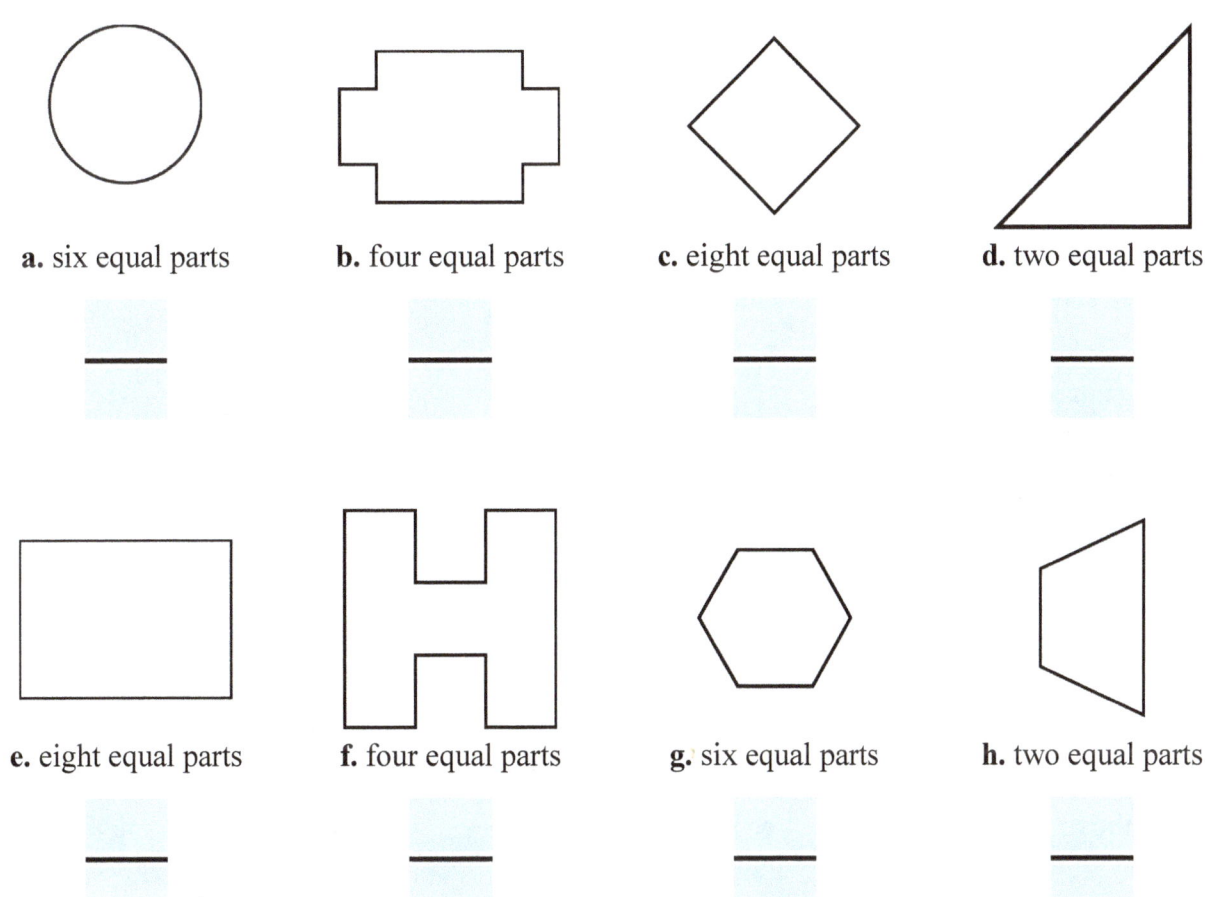

a. six equal parts **b.** four equal parts **c.** eight equal parts **d.** two equal parts

e. eight equal parts **f.** four equal parts **g.** six equal parts **h.** two equal parts

Puzzle Corner

Is it true that 1/8 of the shape is shaded? Explain.

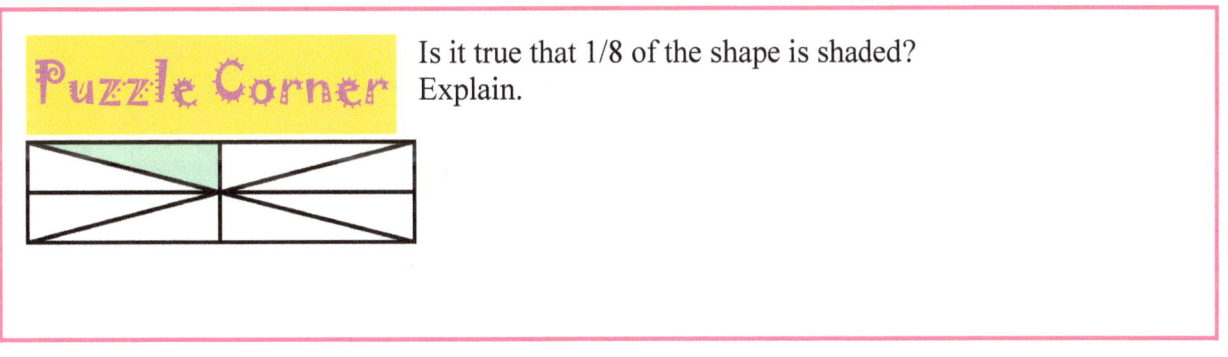

Understanding Fractions 2

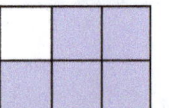 $\frac{5}{6}$

Here, the rectangle is one whole. It is divided into six equal parts, and five of those are colored. Those five parts are **5 sixths**.

 $\frac{4}{4}$

Now the whole has four equal parts. All four parts are colored. We have 4 fourths or one whole.

 $\frac{3}{8}$

"three eighths"

The top number tells **how many** (colored) **PARTS** we have. It is the **numerator** — it enumerates or gives the count of the parts.

The bottom number tells **how many equal parts the whole is divided into**. It is the **denominator** — it denominates or names the type of parts we have.

After halves, we use ordinal numbers to name the fractional parts (thirds, fourths, fifths, sixths, sevenths, eighths, tenths, and so on).

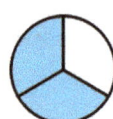 $\frac{2}{3}$

2 thirds

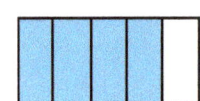 $\frac{4}{5}$

4 fifths

 $\frac{10}{10}$

10 tenths

1. Color parts to illustrate each fraction.

   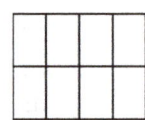  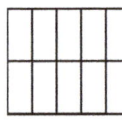

a. 2 thirds   **b.** 4 sixths   **c.** 3 fifths   **d.** $\frac{3}{8}$ **e.** $\frac{2}{2}$ **f.** $\frac{9}{10}$

2. Write each fraction, and read it aloud.

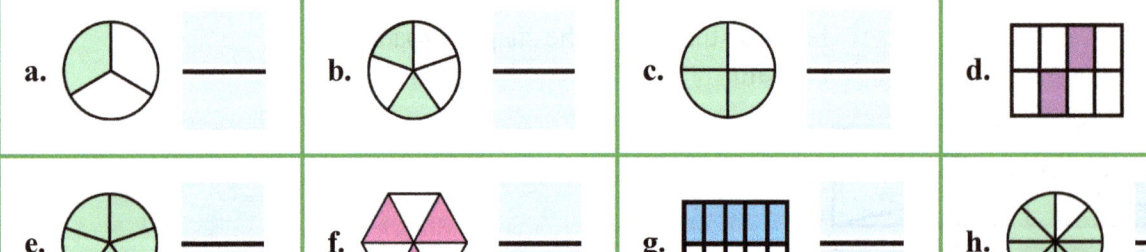

How to draw pie models		
Halves: 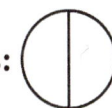 Split the circle with a straight line.	**Thirds:**  Draw lines at 12 o'clock, 4 o'clock, and 8 o'clock.	**Fourths:**  First draw halves, then split those like a cross pattern.
Fifths: 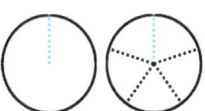 Draw as if a man doing jumping jacks.	**Sixths:**  First draw thirds, then split those.	**Eighths:**  First draw fourths, then split those.

3. Draw the pie models and color the parts to illustrate the fractions.

a. $\dfrac{2}{3}$ b. $\dfrac{2}{5}$ c. $\dfrac{3}{6}$ d. $\dfrac{6}{8}$

e. $\dfrac{4}{5}$ f. $\dfrac{3}{8}$ g. $\dfrac{3}{3}$ h. $\dfrac{7}{8}$

4. Divide the shapes into equal parts, and color some of the parts, to show the fractions.

a. $\dfrac{1}{2}$ b. $\dfrac{3}{6}$ c. $\dfrac{1}{3}$ d. $\dfrac{3}{4}$

5. Color in the whole shape = 1 whole. Then write 1 whole as a fraction.

a. 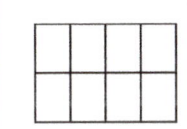 1 = $\dfrac{8}{8}$ b.  1 = ———   c. 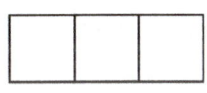 1 = ———   d. 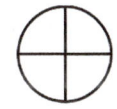 1 = ———   e. 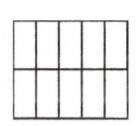 1 = ———

 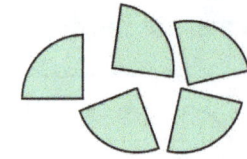 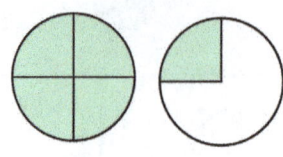

one whole   5 fourths   5 fourths = $\frac{5}{4}$

Here, the circle is one whole. The little pieces are **fourths** (compared to the whole). We have **5 fourths**. If we place these fourths inside the one-whole circles, we get one full circle and another one with one fourth filled in.

How many fourths would you need to have **two wholes**? _____ fourths

6. Color in the parts to show each fraction. In each case, one circle, one hexagon, or one rectangle depicts one whole.

 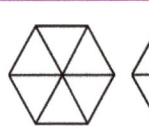 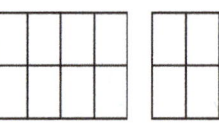 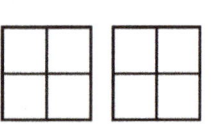

a. $\frac{5}{3}$ b. $\frac{10}{6}$ c. $\frac{11}{8}$ d. $\frac{8}{4}$

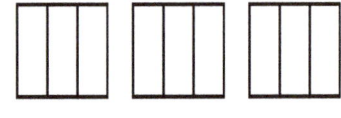 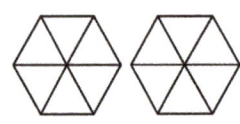 

e. $\frac{7}{3}$ f. $\frac{8}{6}$ g. $\frac{7}{2}$

h. $\frac{19}{4}$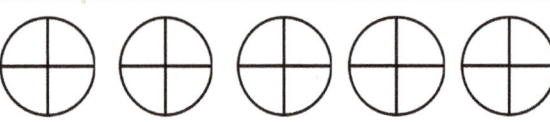

7. <u>Counting activity.</u> We can count in fractions just like counting with whole numbers. For example, counting in sixths, we count:

1 sixth, 2 sixths, 3 sixths, 4 sixths, 5 sixths, <u>6 sixths</u> which also equals <u>one whole</u>.
7 sixths, 8 sixths, 9 sixths, 10 sixths, 11 sixths, <u>12 sixths</u> = <u>two wholes</u>. And so on.

Count in fractions with your teacher or in a circle of students. Each person says the next count. Every time you come to a whole number, name it. If someone gets stuck, others can help.

a. Count in halves, up to four wholes.   **b.** Count in thirds, up to four wholes.

c. Count in fourths, up to four wholes.   **d.** Count in fifths, up to four wholes.

8. Write the fraction and say it aloud.

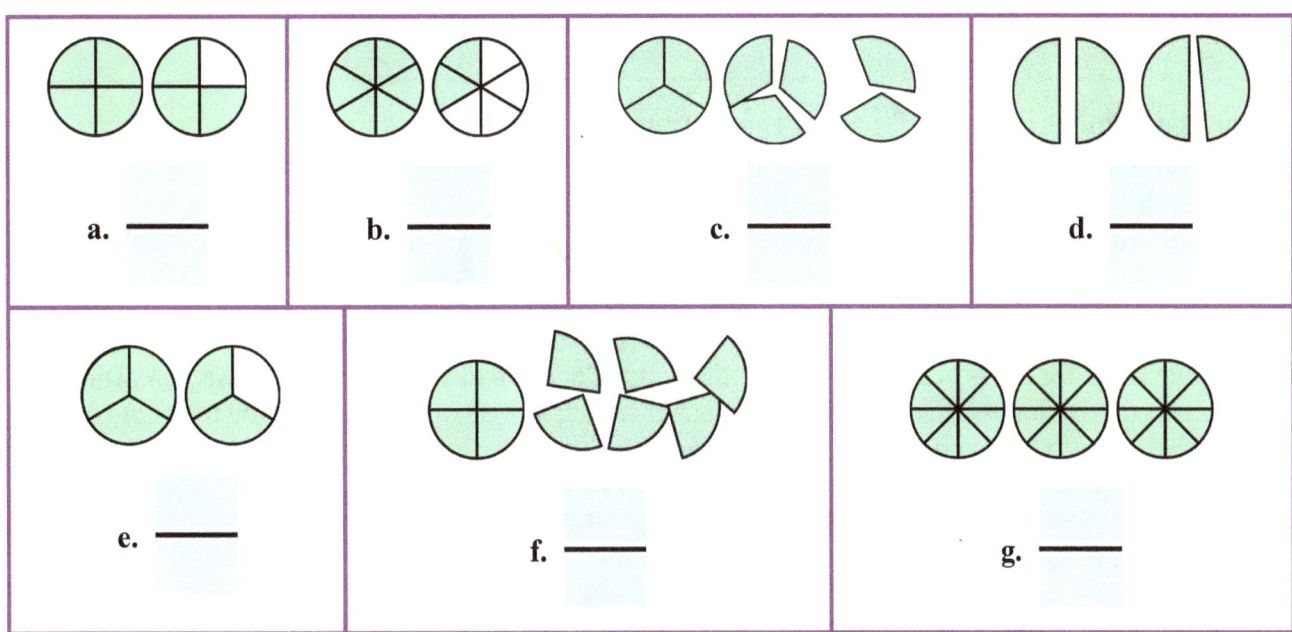

a. ____   b. ____   c. ____   d. ____

e. ____   f. ____   g. ____

9. The picture shows a fractional part. Draw a one whole that corresponds to it. Are there several ways to draw it?

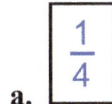

a.

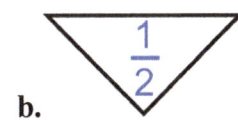

b.   c.

Puzzle Corner   The picture shows fractional parts. Draw a one whole that corresponds to it.

a. $\frac{2}{5}$

b. $\frac{3}{6}$

c. $\frac{5}{8}$

d. $\frac{3}{2}$

Fractions on the Number Line 1

This is a number line from 0 to 1. It is partitioned or divided into *six* parts. Each sixth, or one part, is this much: ├─┤. The green rectangle shows the two parts out of six parts in total.

The point marks the fraction $\frac{2}{6}$ on the number line.

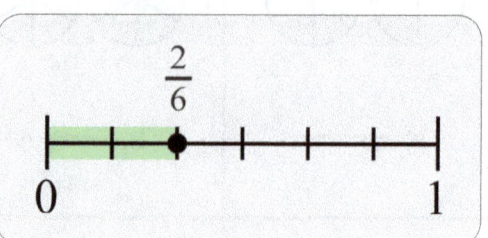

Note: the point for 2/6 is at the right end of the rectangle. Normally, we simply draw a *point* (not a rectangle) on the number line to indicate a number (and fractions *are* numbers). The rectangle is drawn here just to help you understand the concept.

Here you see all the fractions from $\frac{0}{6}$ to $\frac{6}{6}$ on the number line. Note that $\frac{0}{6}$ is the same as 0, and $\frac{6}{6}$ is the same as 1.

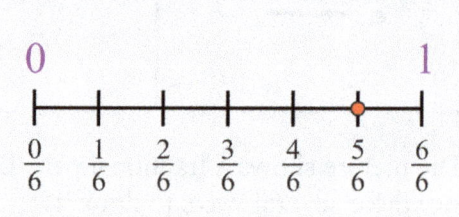

The fraction $\frac{5}{6}$ is marked with a dot. Imagine or draw a rectangle extending from zero to 5/6. The length of that rectangle would be five times ├─┤.

1. Write the fractions under every tick mark, including under 0 and 1.

a.

b.

2. Write the fraction marked by the dot on the number line.

a.

b.

c.

d.

3. **a.** Which one of these number lines is divided into sixths?

   b. Mark the fraction 2/6 on it.

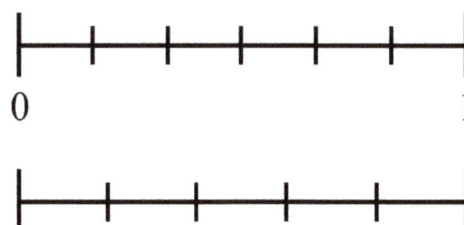

4. Mark the given fraction on the number line (with a dot).

a.

$\frac{2}{4}$

b.

$\frac{3}{5}$

c.

$\frac{4}{6}$

d.

$\frac{1}{8}$

5. Write the fraction marked by the big dot on the number line.

a.

b.

c.

d.

6. James says, incorrectly, that the dot marks the fraction 4/9.

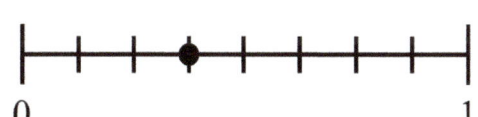

   a. What fraction does the dot mark in reality?

   b. Explain what James could be doing wrong.

7. Mark the given fraction on the number line. (First, you need to partition the number line into equal parts.)

a.
$\frac{1}{4}$

b.
$\frac{1}{3}$

c.
$\frac{3}{4}$

d.
$\frac{1}{6}$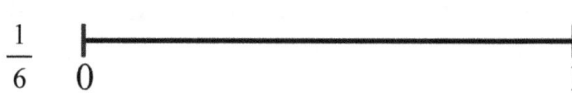

e.
$\frac{5}{6}$

f.
$\frac{3}{8}$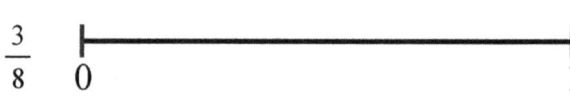

g.
$\frac{4}{5}$

h.
$\frac{5}{8}$

8. Mark the fractions $\frac{5}{6}$, $\frac{5}{8}$, $\frac{5}{5}$, and $\frac{5}{10}$ on the number lines.

   Which is the biggest fraction?

   Which is the smallest?

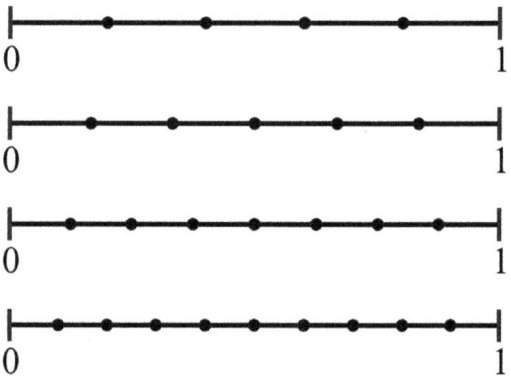

9. Eric and Erika marked the fraction 1/3 in the illustration in different places.

   Which way is correct? Why?

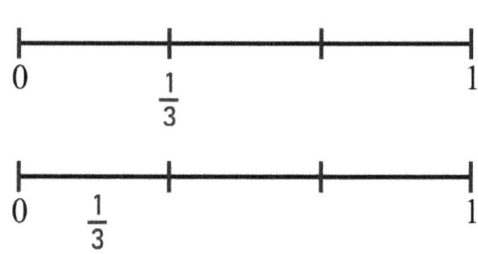

Fractions on the Number Line 2

Here, the interval from 0 to 1 on the number line is divided into **four** parts. The interval from 1 to 2 is also divided into **four** parts, and similarly from 2 to 3.

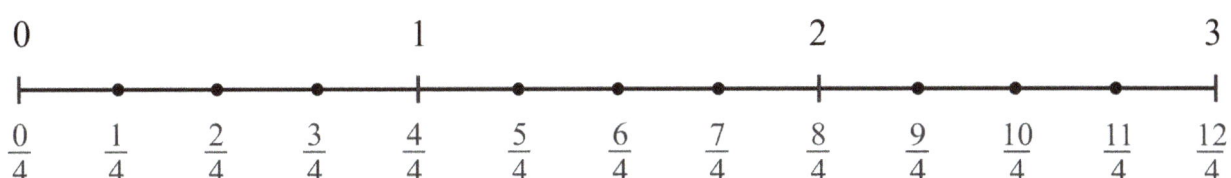

We can count in fourths, starting with zero fourths, then 1 fourth, 2 fourths, and so on.

Notice especially: $1 = \dfrac{4}{4}$, $2 = \dfrac{8}{4}$, and $3 = \dfrac{12}{4}$.

1. Mark the fractions on the number lines. <u>Note:</u> some of them are whole numbers!

   a. $\dfrac{7}{6}, \dfrac{11}{6}, \dfrac{18}{6}, \dfrac{3}{6}, \dfrac{13}{6}$

   b. $\dfrac{6}{5}, \dfrac{11}{5}, \dfrac{15}{5}, \dfrac{9}{5}, \dfrac{13}{5}$

   c. $\dfrac{12}{8}, \dfrac{17}{8}, \dfrac{21}{8}, \dfrac{5}{8}, \dfrac{16}{8}$

   d. Write the whole number 2 as a fraction using sixths. (One of the number lines can help.)

   e. Write the whole number 3 as a fraction using eighths.

2. Divide each interval from one whole number to the next into three parts. Then mark these fractions on the number line: $\frac{2}{3}, \frac{5}{3}, \frac{9}{3}, \frac{8}{3}, \frac{4}{3}$.

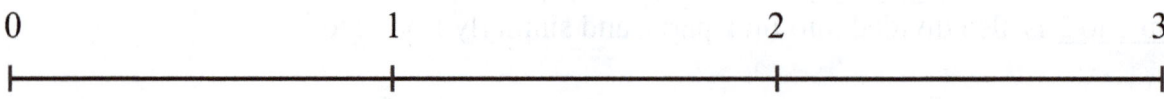

3. Write 4 as a fraction using thirds. (You can imagine extending the above number line.)

4. Write these whole numbers as fractions. You can use pie pictures or the number lines from previous exercises to help. Try to find a **shortcut** or a **general principle** that will allow you to write a whole number as a fraction with a given denominator.

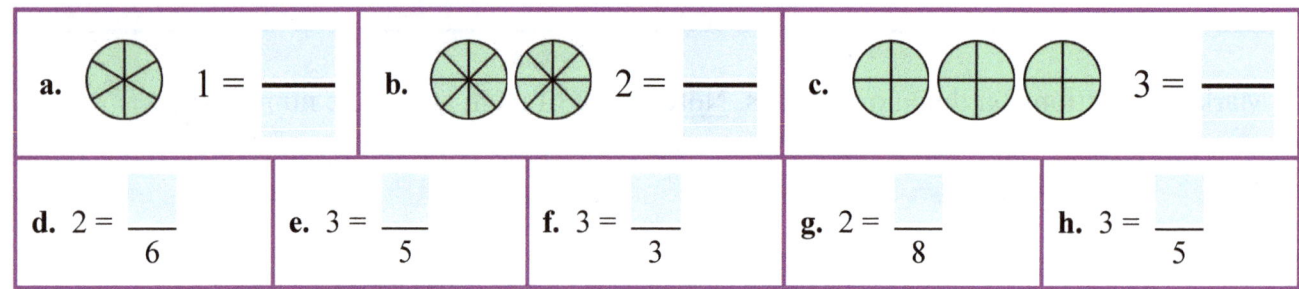

a. 1 = ——   b. 2 = ——   c. 3 = ——

d. $2 = \frac{}{6}$ e. $3 = \frac{}{5}$ f. $3 = \frac{}{3}$ g. $2 = \frac{}{8}$ h. $3 = \frac{}{5}$

5. Did you find a shortcut? Check with your teacher or on the next page.
   Also, explain how to write 5 as a fraction, using eighths, and also, using thirds.

6. Mark these fractions on the number line: $\frac{7}{3}, \frac{14}{3}, \frac{10}{3}, \frac{12}{3}, \frac{9}{3}$.

7. Partition the number line below into halves. Then mark a fraction for each tick mark, including for the whole numbers.

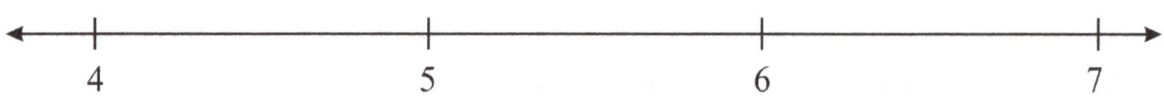

To write a whole number as a fraction, you can use multiplication.

Example: Write 4 as a fraction, using sixths.

Each whole is six sixths. So, 4 wholes is 4 × 6 sixths, or 24 sixths.

In other words, $4 = \frac{24}{6}$.

We just saw that $\frac{24}{6} = 4$. This is the same as the division 24 ÷ 6 = 4! In fact, the fraction line works as a division symbol! Similarly, $\frac{40}{8}$ equals 40 ÷ 8, so, it is 5.

8. Write the whole numbers as fractions.

a. 2 = ——   b. 3 = ——   c. 1 = ——

d. 4 = ——   e. 4 = ——

9. Divide the pies into parts, and color the pies. Write the whole numbers as fractions.

a. $4 = \frac{}{2}$ b. $3 = \frac{}{8}$

c. $4 = \frac{}{4}$ d. $3 = \frac{}{6}$

10. Find three different ways to write number 2 as a fraction.

11. These fractions are actually whole numbers! Which ones?

a. $\frac{6}{6} =$	b. $\frac{21}{7} =$	c. $\frac{24}{6} =$	d. $\frac{20}{2} =$
e. $\frac{20}{4} =$	f. $\frac{8}{8} =$	g. $\frac{12}{3} =$	h. $\frac{30}{5} =$

12. Mark these fractions on the number line: $\frac{19}{4}, \frac{13}{4}, \frac{21}{4}, \frac{24}{4}, \frac{18}{4}$.

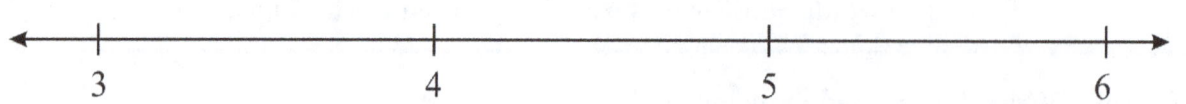

13. Mark these fractions on the number line: $\frac{30}{6}, \frac{26}{6}, \frac{35}{6}, \frac{22}{6}, \frac{31}{6}$.

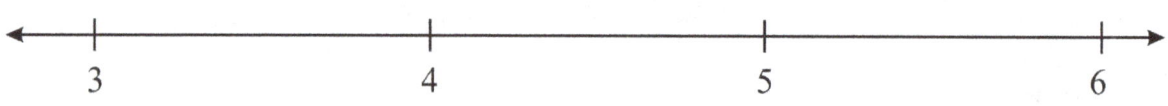

14. Mark these fractions on the number line: $\frac{37}{8}, \frac{40}{8}, \frac{39}{8}, \frac{42}{8}, \frac{47}{8}$.

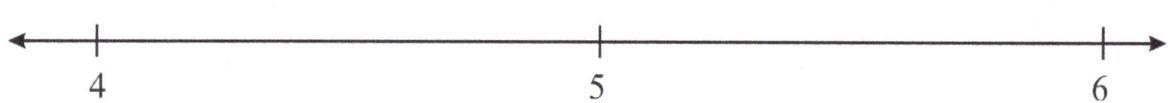

15. Joanna says that the fraction $\frac{69}{8}$ is between 6 and 7. Bill says it is between 7 and 8. Who is right? Explain.

16. Which is the greatest number? $\frac{40}{5}, \frac{40}{8}, \frac{36}{6}, \frac{36}{4}, \frac{36}{3}$

Mark these numbers on the number line: $\frac{5}{4}, \frac{1}{2}, 2, \frac{5}{2}, \frac{9}{4}$.　　**Puzzle Corner**

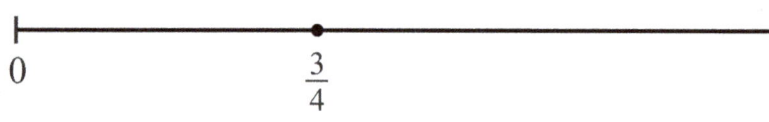

One Whole and Its Fractional Parts

A fraction always relates to some kind of *one whole*. Study the examples below:

 Let's say the one whole is this square. It is divided into 12 parts.
Each part is $\frac{1}{12}$ of the whole. Also, we can write $1 = \frac{12}{12}$.

Maybe the one whole is this line, and $\frac{3}{10}$ of it is colored.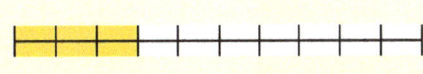

Maybe the one whole is Daddy's salary. To find 5/6 of it, imagine dividing the salary into 6 parts, and taking five of those parts. All six parts form the one whole, or $\frac{6}{6} = 1$

$\frac{7}{12}$ The top number is the **numerator**. It *numerates* or counts *how many pieces* there are.
The bottom number is the **denominator**. It *denominates* or *names* what kind of parts they are.

1. Color parts. Write the colored part *and* the white (uncolored) part as a fraction.

a. Color 1 part.	b. Color 5 parts.	c. Color 8 parts.	d. Color 3 parts.
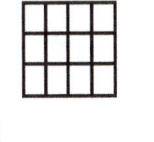 $\frac{1}{12}$ and ___	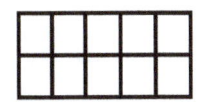 and	 and	 and

2. Color and write one whole as a fraction.

a. 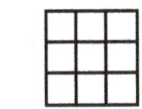 1 = ___	b. 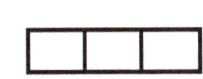 1 = ___	c.  1 = ___	d.  1 = ___	e.  1 = ___

3. Solve.

a. The Jacksons ate $\frac{3}{4}$ of the pie. How much is left?	b. Jerry ate $\frac{1}{6}$ of the pizza. How much is left?

c. Five boys shared a chocolate bar equally. Each one got ___ of the bar.

To show 3/7 on a number line, each whole-number interval (from 0 to 1, from 1 to 2, from 2 to 3, and so on) is divided into seven parts. Three of those parts are colored to show 3/7.

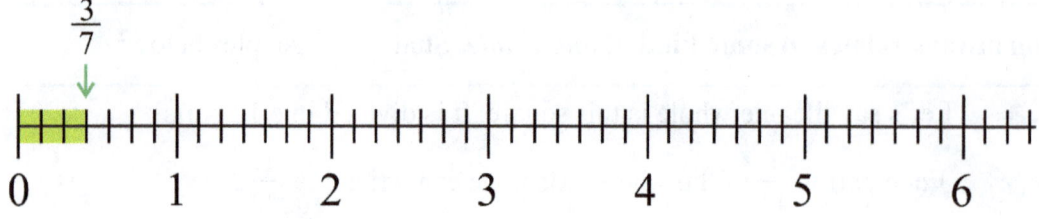

In a **mixed number**, we have a whole number and a fraction. The number line below shows 2 2/7 (two and two sevenths).

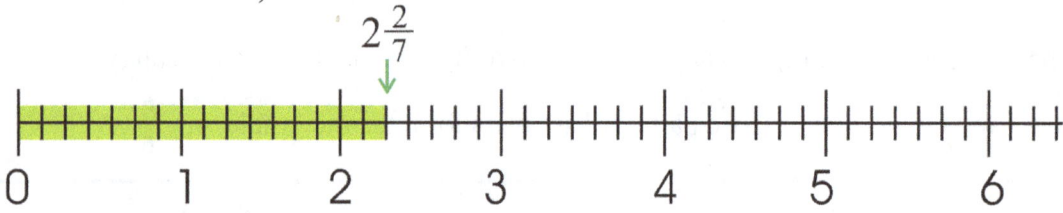

In problems 4 - 6, write the fractions and mixed numbers that the arrows mark on the number line.

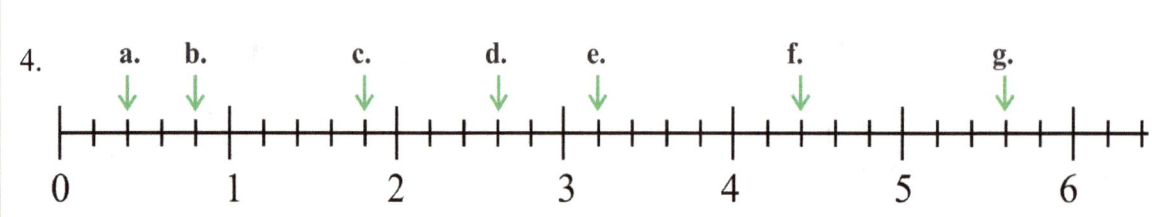

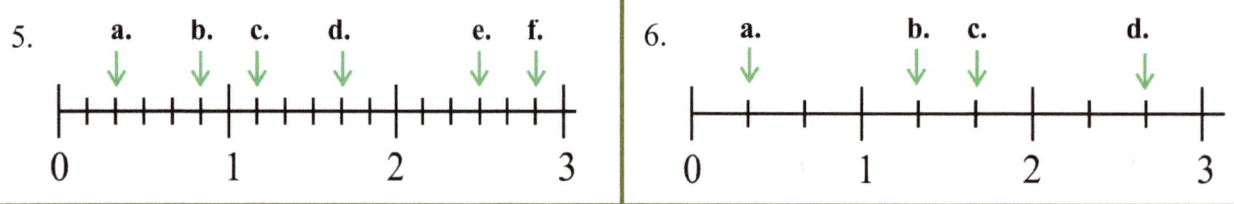

7. Mark the mixed numbers on the number line:
   a. 1 2/4 **b.** 3/4 **c.** 4 1/4 **d.** 5 1/2 **e.** 3 1/4 **f.** 2 3/4
   Hint: First divide each whole-number interval into four parts (using three tick marks).

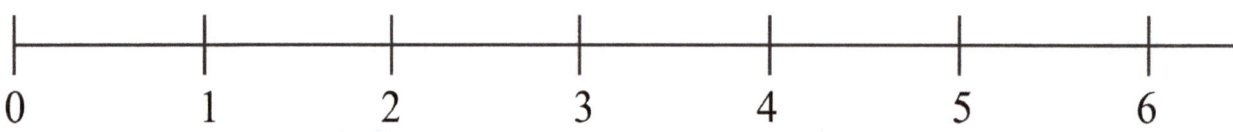

8. Color. Then write an addition, adding the colored and white parts. Notice what sum you get.

a. Color 1 part. 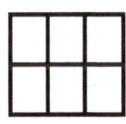 $\frac{1}{6} + \underline{} = 1$	b. Color 10 parts. 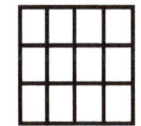	c. Color 3 parts. 	d. Color 15 parts. 

9. Find what fraction is missing from one whole.

a. $\frac{3}{4} + \underline{} = 1$ b. $\frac{6}{7} + \underline{} = 1$ c. $\frac{1}{8} + \underline{} = 1$ d. $\frac{11}{12} + \underline{} = 1$

10. **a.** Mary poured $\frac{1}{4}$ liter of juice from a 1-liter pitcher, and her brother poured another $\frac{1}{4}$ liter. How much juice is left in the pitcher?

b. A loaf of bread was cut into 20 slices. Jack and John ate three slices each. What fractional part of the bread is left?

11. Let's review how to find a fractional part using division.

a. Remember division? Find 1/10 of 90 km.

Then find 4/10 of 90 km.

b. A restaurant bill was $45.50. It was divided so that Cindy paid 2/5 of it and Sandy paid 3/5 of it.

How many dollars did Cindy pay?

How many dollars did Sandy pay?

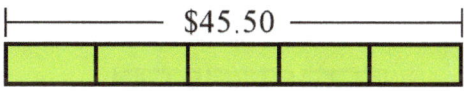

c. Dad used 2/9 of his $2,700 paycheck.

What fractional part is left of his paycheck?

How many dollars are left of his paycheck?

Mixed Numbers

Mixed numbers have two parts: a whole-number part and a fractional part.

This picture illustrates $3\frac{5}{8}$: *three and five eighths*.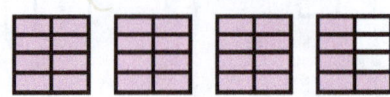

Notice: the colored portion is $3\frac{5}{8}$. The uncolored part is $\frac{3}{8}$.

If we add the colored and uncolored parts, we get four wholes: $3\frac{5}{8} + \frac{3}{8} = 4$.

1. Write the mixed numbers these pictures illustrate.

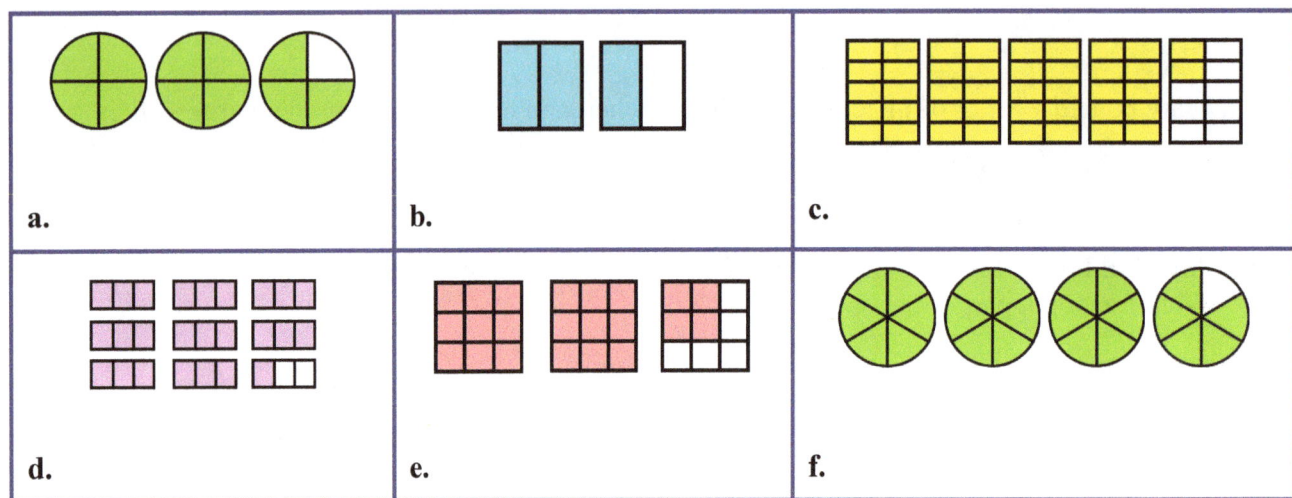

2. Write an addition sentence, adding what is colored and what is not. Look at the example.

a. $2\frac{2}{4} + \frac{2}{4} = 3$

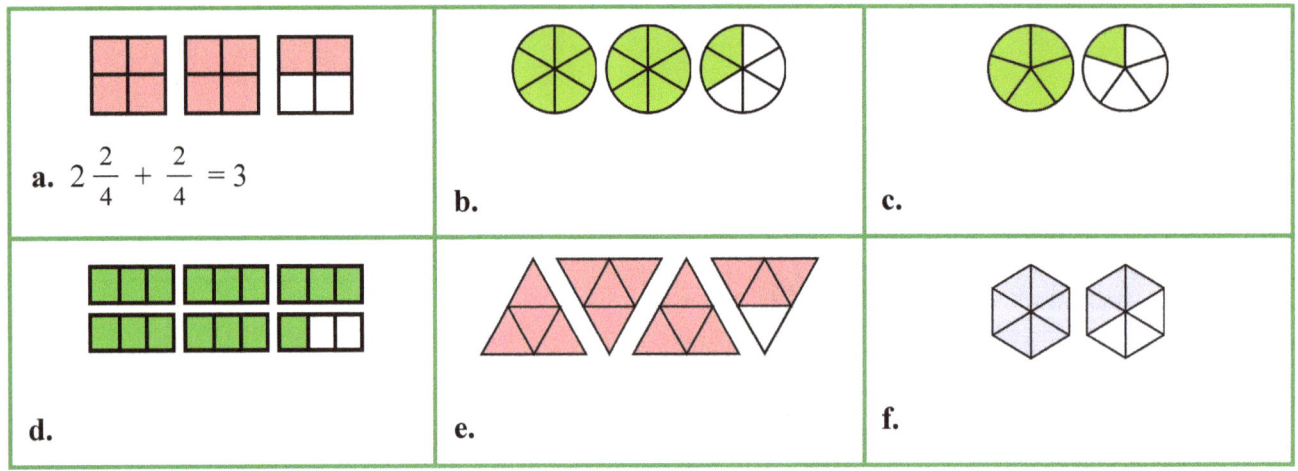

3. How much is missing from the next whole number?

a. $1\frac{1}{4} + \underline{} = 2$ b. $3\frac{2}{10} + \underline{} = 4$ c. $8\frac{4}{9} + \underline{} = 9$ d. $5\frac{1}{8} + \underline{} = 6$

Whole numbers as fractions

The pies show the whole number 3 using sixth parts. To write the number 3 as a fraction we figure out how many sixths there are in total.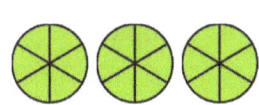

Since one pie has 6 sixths, three whole pies have 3 × 6 = 18 sixths. So, $3 = \frac{18}{6}$.

Example 1. Find the missing number: $2 = \frac{\square}{7}$. In other words, write the whole number 2 as a fraction, using sevenths. Two whole pies have 2 × 7 = 14 sevenths. So $2 = \frac{14}{7}$.

Example 2. The simplest way to write any whole number as a fraction is to use the denominator 1. For example, $23 = \frac{23}{1}$ and $8 = \frac{8}{1}$.

Example 3. Since 45 is divisible by 5, the fraction $\frac{45}{5}$ is equal to the whole number 9.

In other words, you can think of the fraction line as <u>a division symbol.</u> Fractions are divisions!

4. Write the whole numbers as fractions. You can color the shapes to help you.

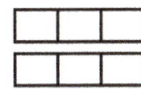 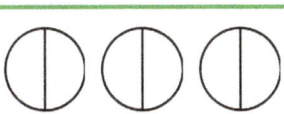 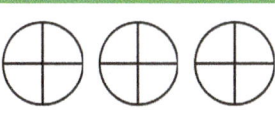

a. $2 = \frac{\square}{\square}$ b. $\_\_\_ = \frac{\square}{\square}$ c. $\_\_\_ = \frac{\square}{\square}$

5. Write whole numbers as fractions.

a. $3 = \frac{\square}{4}$	b. $1 = \frac{\square}{9}$	c. $4 = \frac{\square}{1}$	d. $7 = \frac{\square}{5}$	e. $6 = \frac{\square}{10}$
f. $7 = \frac{\square}{1}$	g. $10 = \frac{\square}{6}$	h. $20 = \frac{\square}{3}$	i. $24 = \frac{\square}{2}$	j. $50 = \frac{\square}{5}$

6. Write these whole numbers as fractions in many ways.

a. $1 = \frac{\square}{2} = \frac{\square}{4} = \frac{\square}{7} = \frac{\square}{9} = \frac{\square}{20}$ b. $4 = \frac{\square}{1} = \frac{\square}{5} = \frac{\square}{10} = \frac{\square}{11} = \frac{\square}{30}$

7. Which whole numbers are these fractions?

a. $\frac{12}{4} =$	b. $\frac{63}{7} =$	c. $\frac{120}{4} =$	d. $\frac{81}{9} =$	e. $\frac{300}{10} =$

We can "break apart" 7/8 in many different ways. From each "breakage" we can write an addition.

For example: 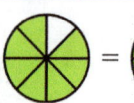 = 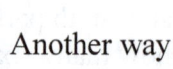 and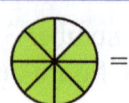

$$\frac{7}{8} = \frac{4}{8} + \frac{1}{8} + \frac{2}{8}$$

Another way: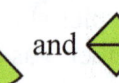

$$\frac{7}{8} = \frac{3}{8} + \frac{2}{8} + \frac{2}{8}$$

8. Write each fraction or mixed number as an addition in different ways.

a.

$$\frac{4}{5} = \boxed{} + \boxed{}$$

$$\frac{4}{5} =$$

$$\frac{4}{5} =$$

b.

$$\frac{5}{8} =$$

$$\frac{5}{8} =$$

$$\frac{5}{8} =$$

c.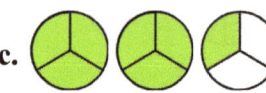

$$2\frac{1}{3} =$$

$$2\frac{1}{3} =$$

$$2\frac{1}{3} =$$

d.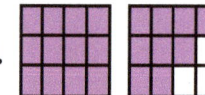

 =

 =

 =

e.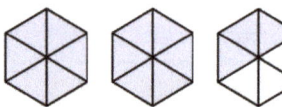

 =    =

 =   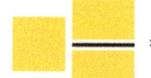 =

9. A pitcher can hold 2 cups when full. Jill poured 1 ¼ cups of water into it. How much more water can she put into the pitcher?

10. A recipe calls for 1 1/3; pounds of beef. Sandra has a 2 pound package of beef. How much beef will be left over?

11. Jerry had two loaves of bread, which he divided into 12 slices each. Then he ate five slices. Write the amount of loaves left as a *mixed number*.
    Hint: Draw a picture to illustrate the bread.

12. One toy car is 1 ½ inches long. If you put three of them end-to-end, how long is your "train" of cars?

13. Emma is making cookies. She is using a ¼-cup measuring cup. She needs 1 ¼ cup of flour. How many scoops of flour will she need using the ¼-cup measuring cup?

Puzzle Corner

Are these additions correct? If not, change something to make the addition equation correct. You can draw pie pictures to help.

a. $3\frac{3}{4} = \frac{7}{4} + \frac{2}{4} + \frac{5}{4}$

b. $3 = \frac{6}{5} + \frac{3}{5} + 1\frac{1}{5}$

Mixed Numbers and Fractions

How to convert fractions to mixed numbers

Example 1. The fraction $\frac{23}{6}$ is more than one whole... so we can write it as a mixed number.

(How do we know it is more than 1? Because 1 = 6/6, and 23 sixths is more than six sixths.)

One way to write $\frac{23}{6}$ as a mixed number is to shade 23 sixths in the pie pictures:

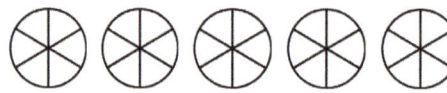

If you do, you will get 3 whole pies, and 5 sixths shaded in one more pie. So the total is 3 5/6.

A quicker way is to use *division*. To find the number of whole pies we get from all those sixths, we need to figure how many 6's are in 23 (since each six slices makes a whole pie). The answer to that is found **by dividing 23 ÷ 6.**

So, we think of the fraction $\frac{23}{6}$ as the division problem 23 ÷ 6. Since 23 ÷ 6 = 3 R5, we get three full pies, and five slices left over. This means that the mixed number we get is $3\frac{5}{6}$.

Example 2. To change $\frac{34}{5}$ into a mixed number, think of the fraction as a division, and <u>divide</u>.

We get 34 ÷ 5 = 6 R4. This means there are six whole pies, and four slices left over, or 6 4/5.

1. Write these fractions as mixed numbers. You can color parts to help.

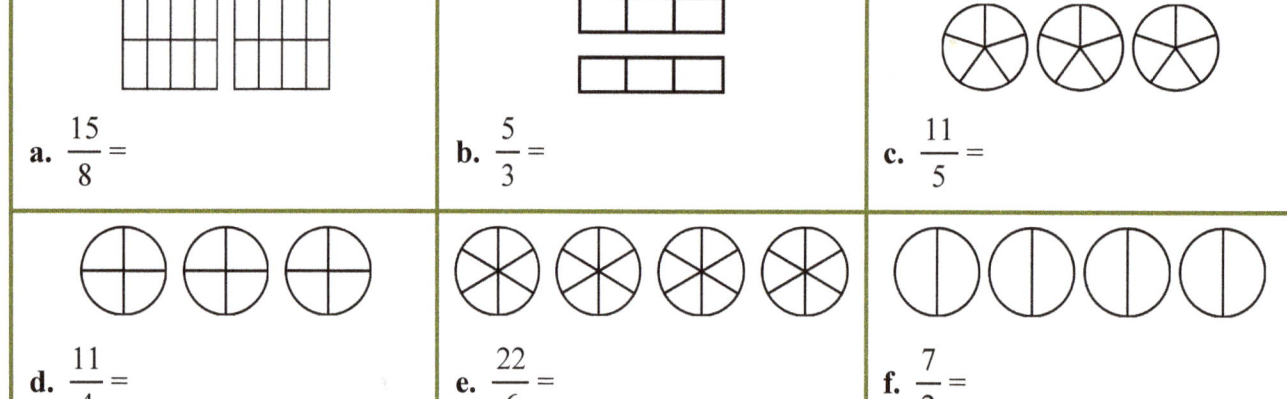

a. $\frac{15}{8} =$

b. $\frac{5}{3} =$

c. $\frac{11}{5} =$

d. $\frac{11}{4} =$

e. $\frac{22}{6} =$

f. $\frac{7}{2} =$

2. Write the fractions as mixed numbers.

a. $\frac{13}{5} =$	b. $\frac{11}{3} =$	c. $\frac{23}{4} =$	d. $\frac{17}{2} =$
e. $\frac{26}{7} =$	f. $\frac{55}{9} =$	g. $\frac{22}{10} =$	h. $\frac{58}{8} =$

How to convert mixed numbers to fractions

Here we see $3\frac{1}{5}$. To write it as a fraction, we need to figure out how many fifths there are in total.

- In the three whole pies there are $3 \times 5 = 15$ fifths.
- Besides those, we have one more fifth.

In total, there are 16 fifths, which gives us our fraction: $3\frac{1}{5} = \frac{16}{5}$.

Example 6. here we have $2\frac{4}{9}$. We multiply $2 \times 9 = 18$ to get the number of ninths in the two wholes. Then we add the 4 ninths, to get a total of 22 ninths. So, $2\frac{4}{9} = \frac{22}{9}$.

Shortcut: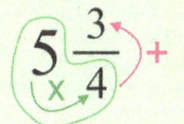

Numerator: $5 \times 4 + 3 = 23$

Denominator: 4

$= \frac{23}{4}$

3. Write the mixed numbers as fractions. You can use the pictures to help you.

a.  $1\frac{4}{9} = $

b. 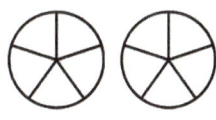 $1\frac{3}{5} = $

c. 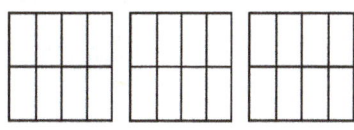 $2\frac{5}{8} = $

4. Write the mixed numbers as fractions.

a. $2\frac{2}{5} = $	b. $1\frac{1}{3} = $	c. $3\frac{1}{4} = $	d. $4\frac{1}{2} = $
e. $5\frac{1}{4} = $	f. $6\frac{1}{3} = $	g. $8\frac{2}{3} = $	h. $8\frac{1}{10} = $

5. Match.

$2\frac{2}{5}$ $1\frac{4}{5}$ $4\frac{7}{8}$ $3\frac{1}{5}$ $3\frac{3}{5}$ $5\frac{5}{8}$ $4\frac{1}{8}$

$\frac{45}{8}$ $\frac{12}{5}$ $\frac{16}{5}$ $\frac{9}{5}$ $\frac{33}{8}$ $\frac{18}{5}$ $\frac{39}{8}$

6. Practice some more. Write the fractions as mixed numbers, and the mixed numbers as fractions.

a. $\frac{45}{6} =$	b. $7\frac{1}{3} =$	c. $\frac{47}{20} =$	d. $3\frac{5}{6} =$
e. $\frac{23}{4} =$	f. $4\frac{2}{5} =$	g. $\frac{34}{7} =$	h. $10\frac{3}{4} =$

7. Write as an addition in different ways. Use at least 3 addends.

a.

b.

8. A library has 2,610 children's books. Of those, 7/9 are fiction.

How many books are fictional?

How many are not?

9. How much is missing from the next whole number?

a. $11\frac{1}{12} + \underline{} = 12$	b. $8\frac{2}{9} + \underline{} = 9$	c. $29\frac{4}{15} + \underline{} = 30$	d. $6\frac{7}{20} + \underline{} = 7$

10. Write whole numbers as fractions.

a. $8 = \frac{}{2}$	b. $10 = \frac{}{7}$	c. $6 = \frac{}{11}$	d. $20 = \frac{}{4}$	e. $12 = \frac{}{8}$

Comparing Fractions 1

1. Color the whole shape. Then write 1 whole as a fraction. Lastly, read what you wrote with numbers.

a. $1 = \dfrac{}{}$ "One whole is 3 thirds."

b. $1 = \dfrac{}{}$

c. $1 = \dfrac{}{}$

d. $1 = \dfrac{}{}$

2. Color. Then compare and write <, >, or = . Which is more "pie" to eat?

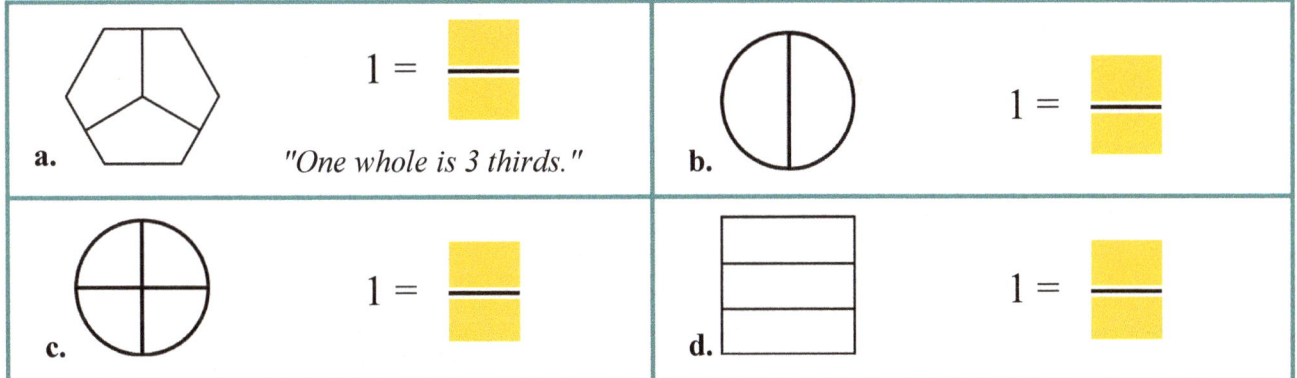

a. $\dfrac{1}{3}$ $\dfrac{1}{2}$

b. $\dfrac{2}{3}$ $\dfrac{3}{4}$

c. 1 whole $\dfrac{3}{4}$

d. $\dfrac{2}{4}$ $\dfrac{1}{2}$

e. 1 whole $\dfrac{2}{2}$

f. $\dfrac{1}{2}$ $\dfrac{2}{3}$

3. Divide the shapes into two, three, or four equal parts so that you can color the fraction. Then compare and write <, >, or = .

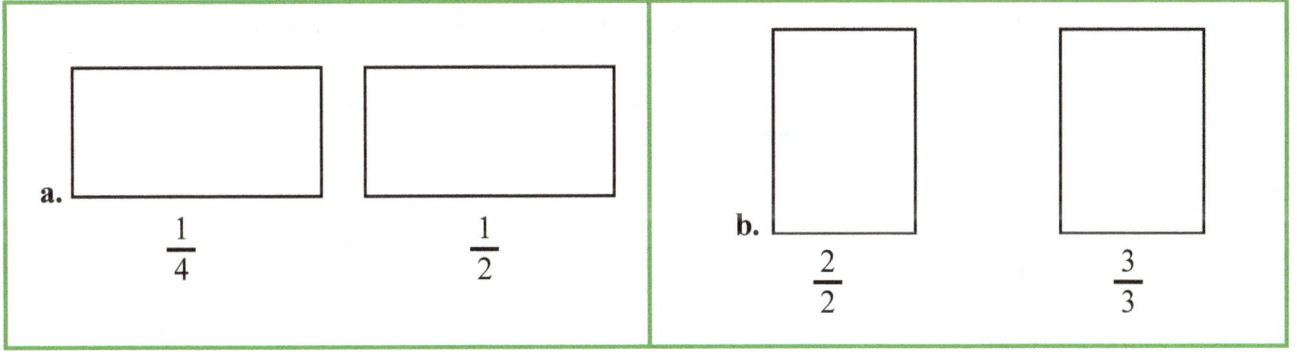

a. $\dfrac{1}{4}$ $\dfrac{1}{2}$

b. $\dfrac{2}{2}$ $\dfrac{3}{3}$

More fractions

When we divide something into FIVE equal parts, the parts are called *fifths*.
When we divide something into SIX equal parts, the parts are called *sixths*.

Here, five-sixths of the square is colored. We write $\frac{5}{6}$ or 5/6. 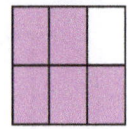	Here, two-fifths of the circle are colored. We write $\frac{2}{5}$ or 2/5. 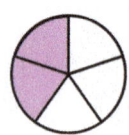

4. Color the given fraction.

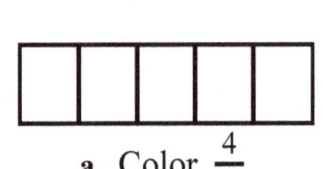

a. Color $\frac{4}{5}$ b. Color $\frac{2}{5}$ c. Color $\frac{5}{6}$ d. Color $\frac{1}{6}$

5. Color. Then compare and write < , > , or = . Which is more "pie" to eat?

a. $\frac{1}{5}$ ___ $\frac{1}{6}$ b. $\frac{3}{4}$ ___ $\frac{3}{5}$ c. $\frac{4}{6}$ ___ $\frac{2}{3}$

d. $\frac{2}{5}$ ___ $\frac{1}{2}$ e. $\frac{3}{6}$ ___ $\frac{1}{2}$ f. $\frac{1}{5}$ ___ $\frac{1}{4}$

6. Divide the shapes into two, three, or four equal parts so that you can color the fraction. Then compare and write < , > , or = .

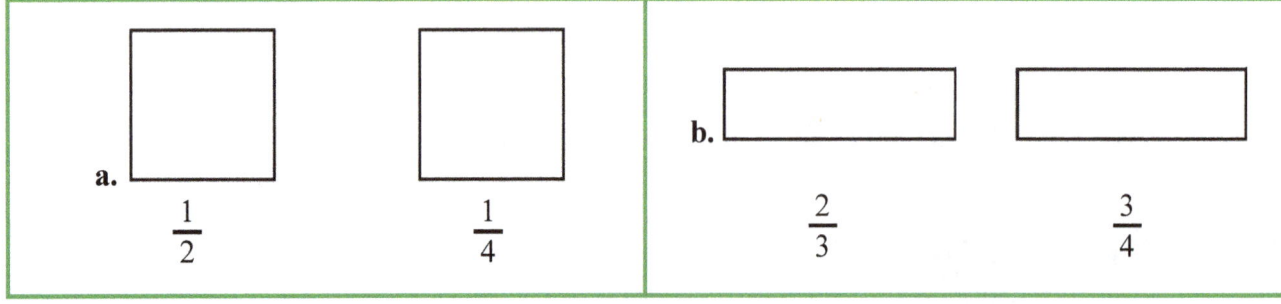

a. $\frac{1}{2}$ ___ $\frac{1}{4}$ b. $\frac{2}{3}$ ___ $\frac{3}{4}$

Comparing Fractions 2

Example 1. In the illustration, each rectangle is one whole. We can see that 2/8 is less than 1/2.

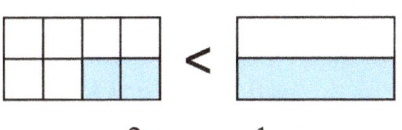

$$\frac{2}{8} < \frac{1}{2}$$

Example 2. With number lines, the point that is furthest from 0 marks the bigger number. And, fractions are numbers, too.

(Imagine a rectangle drawn from 0 to the given fraction. Which rectangle will be longer?)

We can see that 3/5 is greater than 2/4.

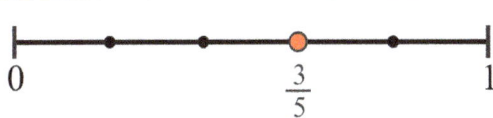

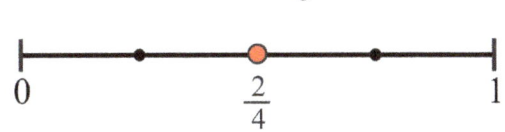

$$\frac{3}{5} > \frac{2}{4}$$

1. Compare and write > or < between the fractions. You can color pieces to help.

a. $\frac{1}{4}$ ☐ $\frac{1}{3}$

b. $\frac{4}{5}$ ☐ $\frac{3}{4}$

c. $\frac{5}{6}$ ☐ $\frac{6}{8}$

2. Compare and write > or < between the fractions.

a. $\frac{1}{5}$ ☐ $\frac{1}{8}$

b. $\frac{2}{5}$ ☐ $\frac{1}{4}$

c. $\frac{1}{3}$ ☐ $\frac{3}{8}$

d. $\frac{2}{3}$ ☐ $\frac{2}{5}$

Example 3. If these were some kind of food, which colored part would give you more to eat?

Obviously, the one on the left. Yet they are both 1/3 of the entire shape! Does this mean that 1/3 is greater than 1/3 sometimes?! What is going on?

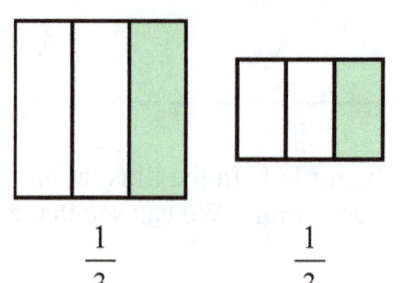

The problem is caused by the fact that <u>the wholes are not the same size</u>. This means we cannot make valid comparisons using fractions. We *can* say that the piece on the left is bigger (in area) than the piece on the right, but we should not do a comparison <u>using fractions</u>.

For a comparison with fractions to be valid, **the fractions have to refer to a whole that is the same size.**

3. Compare the fractions, writing > or < between them, but only if the comparison is valid. If the comparison would not be valid, cross out the entire problem.

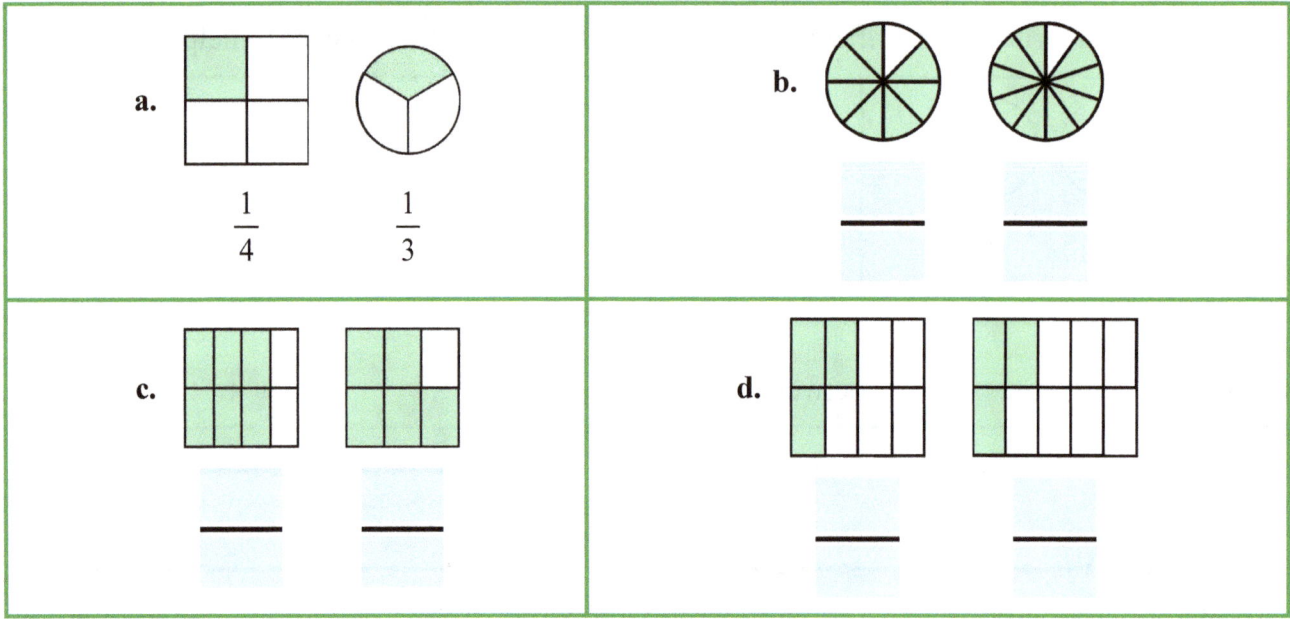

4. Finn says that these fractions are equal, because both 6/6 and 4/4 are equal to 1.

   Hazel says that is not true, because one of them has "more to eat", because it's bigger.

   What do you think?

5. Compare and write > or < between the fractions. If the comparison cannot be made in a valid way, cross the entire problem out.

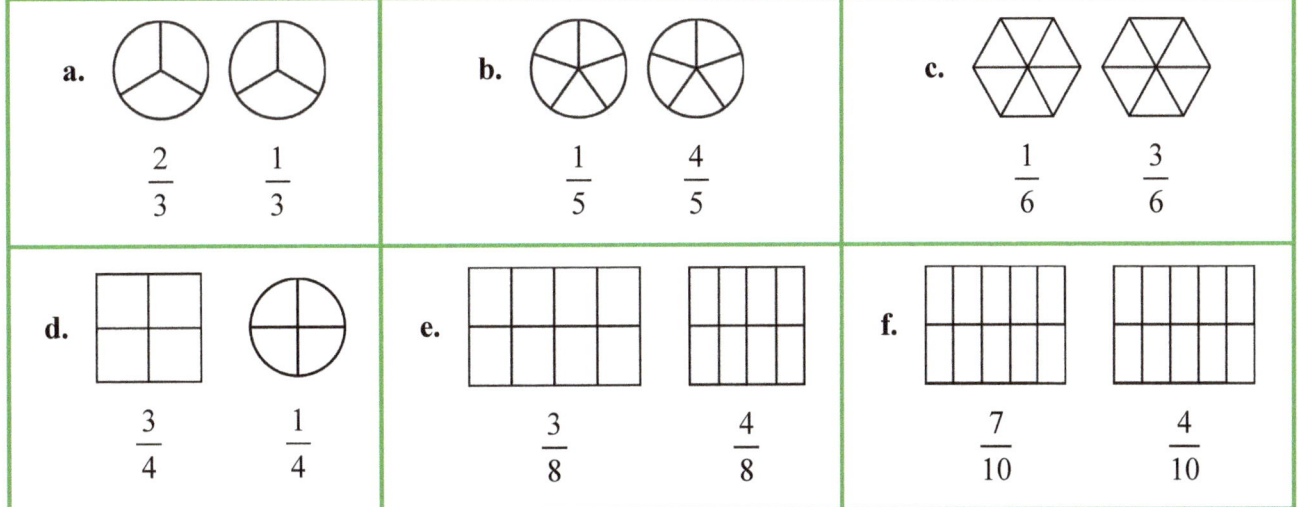

a. $\frac{2}{3}$ $\frac{1}{3}$	b. $\frac{1}{5}$ $\frac{4}{5}$	c. $\frac{1}{6}$ $\frac{3}{6}$
d. $\frac{3}{4}$ $\frac{1}{4}$	e. $\frac{3}{8}$ $\frac{4}{8}$	f. $\frac{7}{10}$ $\frac{4}{10}$

6. Now look at the *valid* comparisons above. Explain how to find the bigger fraction if two fractions have the same denominator (they have the same kind of parts)? For example, how do you tell which is the bigger fraction, 5/8 or 3/8? What about 7/6 or 1/6?

_____

_____

7. Compare, writing > or < between the fractions.

a. $\frac{3}{4}$ ☐ $\frac{5}{4}$	b. $\frac{3}{6}$ ☐ $\frac{1}{6}$	c. $\frac{9}{8}$ ☐ $\frac{8}{8}$	d. $\frac{5}{5}$ ☐ $\frac{2}{5}$
e. $\frac{13}{10}$ ☐ $\frac{3}{10}$	f. $\frac{3}{3}$ ☐ $\frac{5}{3}$	g. $\frac{3}{6}$ ☐ $\frac{6}{6}$	h. $\frac{5}{2}$ ☐ $\frac{2}{2}$

8. These two pitchers are both 1/4 full. Do they have the same amount of water? Explain.

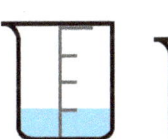

9. Who got to eat more pie? Can you tell?

Nora   Elliott

Comparing Fractions 3

1. Color **one piece** in each pie model, to illustrate **unit fractions**. Then compare the given unit fractions. Write < or > in the box between the fractions.

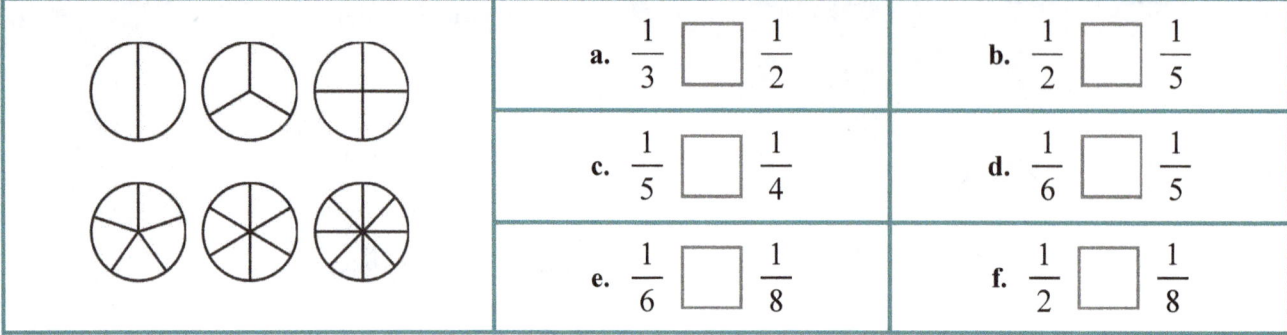

a. $\frac{1}{3}$ ☐ $\frac{1}{2}$	b. $\frac{1}{2}$ ☐ $\frac{1}{5}$
c. $\frac{1}{5}$ ☐ $\frac{1}{4}$	d. $\frac{1}{6}$ ☐ $\frac{1}{5}$
e. $\frac{1}{6}$ ☐ $\frac{1}{8}$	f. $\frac{1}{2}$ ☐ $\frac{1}{8}$

2. **a.** Which is a greater fraction, $\frac{1}{9}$ or $\frac{1}{8}$?

   Explain how you can know that.

   b. When it comes to unit fractions (such as 1/5 and 1/8), how does the size of the denominator (the bottom number) relate to the size of the fraction?

3. Which is greater, $\frac{1}{3}$ or $\frac{1}{4}$?

   Show it using the two number lines.

4. These number lines seem to show that $\frac{1}{5} = \frac{1}{4}$.

   Is that true? Why or why not?
   You can draw a picture to prove your point.

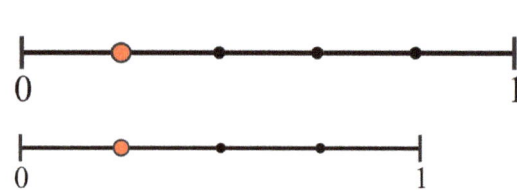

5. **a.** Color these fractions in the fraction bars.

$$\frac{2}{5} \qquad \frac{2}{10} \qquad \frac{2}{2} \qquad \frac{2}{4}$$

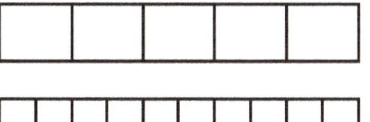

b. What is the same about these fractions?

What is different?

c. When you look at the size of each fraction compared to the others, what do you notice?

6. Compare and write > or < between the fractions.

a. $\frac{2}{6} \qquad \frac{2}{3}$	**b.** 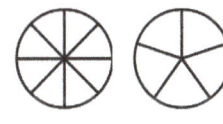 $\frac{4}{8} \qquad \frac{4}{5}$	**c.** 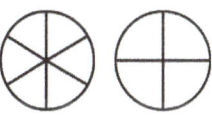 $\frac{3}{6} \qquad \frac{3}{4}$
d. $\frac{5}{6} \qquad \frac{5}{8}$	**e.**  $\frac{2}{2} \qquad \frac{2}{3}$	**f.**  $\frac{7}{12} \qquad \frac{7}{9}$

7. Look at the comparisons above. Notice what is the same about the two fractions in each case, and what is different.

Explain how to tell which fraction is greater if two fractions have the same amount of pieces (the numerators are the same).

For example, how can you tell which is greater, 5/8 or 5/9?

Or, how can you tell which is greater, 6/10 or 6/8?

8. Compare and write >, <, or = between the fractions. You can use the pie pictures to help.

a. $\frac{2}{5}\ \square\ \frac{2}{8}$	b. $\frac{3}{8}\ \square\ \frac{3}{6}$	c. $\frac{1}{2}\ \square\ \frac{1}{5}$	d. $\frac{6}{6}\ \square\ \frac{5}{6}$
e. $\frac{1}{4}\ \square\ \frac{2}{8}$	f. $\frac{3}{8}\ \square\ \frac{5}{8}$	g. $\frac{1}{2}\ \square\ \frac{5}{10}$	h. $\frac{4}{4}\ \square\ \frac{2}{2}$

9. Compare and write >, <, or = between the fractions.

a. $\frac{5}{6}\ \square\ \frac{4}{6}$ b. $\frac{7}{5}\ \square\ \frac{7}{10}$ c. $\frac{9}{10}\ \square\ \frac{5}{10}$ d. $\frac{3}{4}\ \square\ \frac{6}{8}$

e. $\frac{3}{5}\ \square\ \frac{3}{4}$ f. $\frac{2}{3}\ \square\ \frac{4}{6}$ g. $\frac{5}{8}\ \square\ \frac{5}{6}$ h. $\frac{5}{8}\ \square\ \frac{9}{8}$

10. Mark says that 5/3 is less than 7/8 because 5 and 3 are smaller numbers than 7 and 8. Help Mark see the error in his thinking.

11. Write these four fractions in order from smallest to greatest. $\frac{1}{6}$ $\frac{1}{3}$ $\frac{1}{9}$ $\frac{1}{5}$

Puzzle Corner   Find the smallest and the largest fraction in each set.

a. $\frac{3}{5}$ $\frac{5}{8}$ $\frac{2}{3}$ $\frac{7}{10}$

b. $\frac{2}{5}$ $\frac{1}{3}$ $\frac{3}{8}$ $\frac{3}{10}$

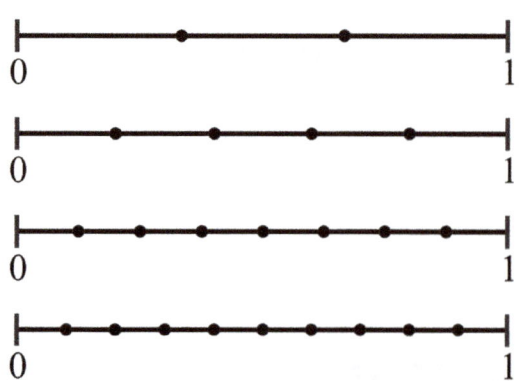

Comparing Fractions 4

1. Compare the fractions, writing >, <, or = between them. If you cannot make a valid comparison, then cross the whole problem out.

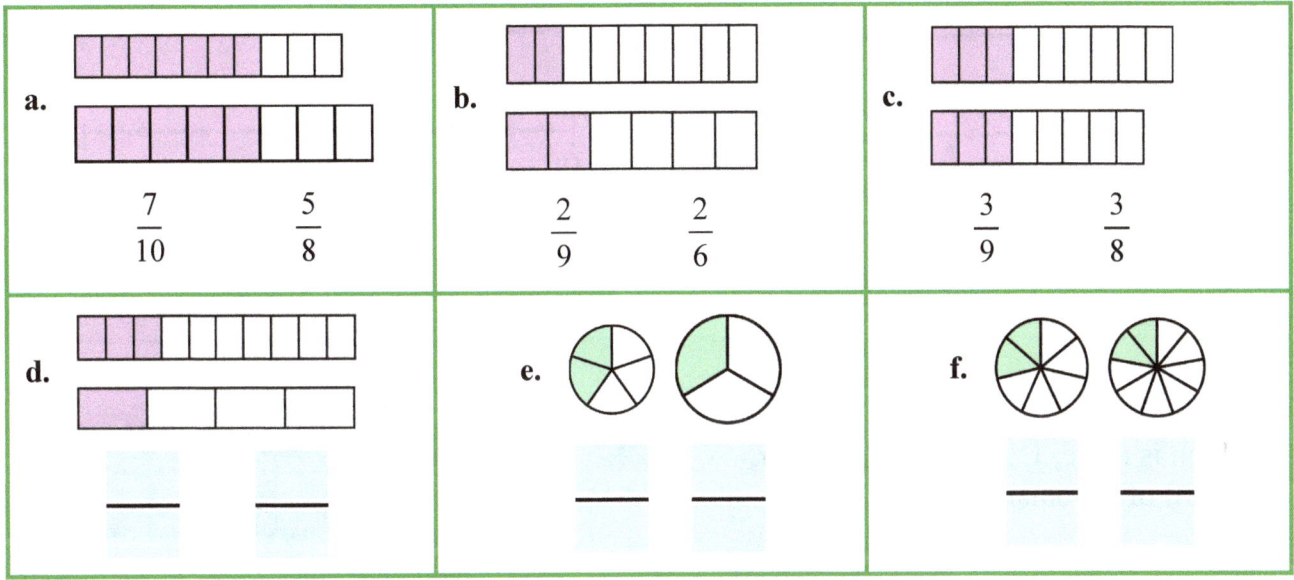

2. Ellie has a blue and a red ribbon that are the same length. She cuts the blue ribbon into 5 equal pieces, and the red ribbon into 4 equal pieces. Which is longer, a piece from the blue ribbon or a piece from the red ribbon?

3. Compare the fractions. Think carefully.

   a. $\dfrac{4}{3} \square \dfrac{3}{3}$ b. $\dfrac{6}{7} \square \dfrac{6}{9}$ c. $\dfrac{9}{10} \square \dfrac{7}{10}$ d. $\dfrac{9}{12} \square \dfrac{9}{5}$

   e. $\dfrac{1}{6} \square \dfrac{1}{4}$ f. $\dfrac{1}{12} \square \dfrac{10}{10}$ g. $\dfrac{3}{8} \square \dfrac{3}{6}$ h. $\dfrac{1}{2} \square \dfrac{8}{8}$

4. These two fractions look like they are the same amount. Yet, 7/8 cannot be equal to 9/10. Or can it?

   Below, you will also see pie pictures of them.

   What do you think? Are they equal?

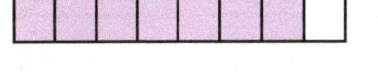

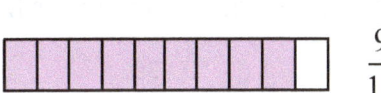

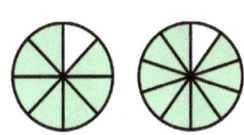

5. Mark the fractions on the number lines. Then find the biggest and the smallest fraction of the three.

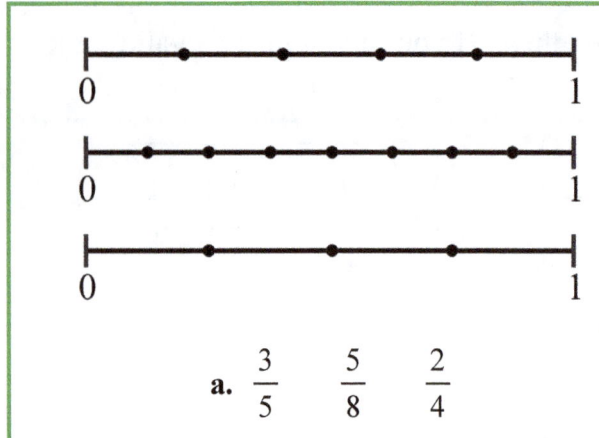

a. $\frac{3}{5}$ $\frac{5}{8}$ $\frac{2}{4}$

b. $\frac{4}{5}$ $\frac{2}{3}$ $\frac{7}{8}$

6. Janet has two bars of dark chocolate. Which is more, 1/12 of the bigger bar, or 2/12 of the smaller bar?

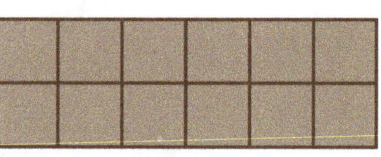

Does that prove that $\frac{2}{12} < \frac{1}{12}$?

7. Six workers bought one large pizza and one small pizza to share evenly. Both pizzas are cut into three equal pieces. Will it be fair if everybody gets one piece (1/3)?

Why or why not?

Write these fractions in order from the least to the greatest:

$\frac{13}{5}, \frac{18}{6}, \frac{8}{4}, \frac{5}{3}, \frac{36}{10}$

Puzzle Corner

Comparing Fractions 5

1. Compare the fractions by writing < or > in the box between them.

If the fractions have the same *kind* of pieces, you can simply compare how many pieces they have.	a.   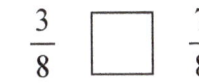	b.   	c. $\dfrac{5}{9}$ ☐ $\dfrac{7}{9}$
			d. $\dfrac{6}{6}$ ☐ $\dfrac{2}{6}$

2. Compare the fractions by writing < or > in the box between them.

If the fractions have the same *amount* of pieces, you can simply compare the size of the pieces.	a.   	b.   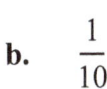	c. $\dfrac{5}{6}$ ☐ $\dfrac{5}{8}$
			d. $\dfrac{2}{6}$ ☐ $\dfrac{2}{5}$

Sometimes one fraction is more than 1/2 and the other is less.

Example 1. Compare $\dfrac{5}{6}$ and $\dfrac{3}{8}$.

Now, 3/8 is less than 1/2. How can you know? Because 4/8 would be exactly 1/2, so 3/8 is less than that. And, 5/6 is more than 1/2. (How do you know?) So, 5/6 > 3/8.

3. Write <, >, or = in the box. <u>Note:</u> Sometimes one of the fractions is actually *equal* to 1/2!

a. $\dfrac{1}{6}$ ☐ $\dfrac{3}{5}$ b. $\dfrac{4}{5}$ ☐ $\dfrac{2}{8}$ c. $\dfrac{3}{4}$ ☐ $\dfrac{2}{5}$ d. $\dfrac{5}{10}$ ☐ $\dfrac{4}{12}$

e. $\dfrac{4}{5}$ ☐ $\dfrac{3}{6}$ f. $\dfrac{1}{9}$ ☐ $\dfrac{2}{3}$ g. $\dfrac{3}{6}$ ☐ $\dfrac{5}{10}$ h. $\dfrac{4}{10}$ ☐ $\dfrac{7}{12}$

4. Write these fractions in order from the smallest to the greatest.

a. $\dfrac{6}{8}, \dfrac{3}{8}, \dfrac{3}{6}$	b. $\dfrac{6}{5}, \dfrac{2}{5}, \dfrac{5}{6}$	c. $\dfrac{1}{4}, \dfrac{1}{7}, \dfrac{5}{8}$

> **Sometimes one fraction is more than 1 whole and the other is less.**
>
> **Example 2.** Compare $\frac{7}{10}$ and $\frac{11}{4}$. Clearly, 11 fourths is more than one since 4/4 makes one.
>
> And, 7/10 is less than 1. So, 7/10 must also be less than 11/4.

5. Compare the fractions. (Write <, >, or = in the box.)

   a. $\frac{8}{7}$ ☐ $\frac{7}{8}$ b. $\frac{9}{12}$ ☐ $\frac{7}{5}$ c. $\frac{3}{4}$ ☐ $\frac{8}{5}$ d. $\frac{11}{12}$ ☐ $\frac{10}{3}$

6. Compare.

a. $\frac{1}{7}$ ☐ $\frac{3}{7}$	b. $\frac{1}{2}$ ☐ $\frac{5}{6}$	c. $\frac{4}{5}$ ☐ $\frac{8}{3}$	d. $1\frac{2}{9}$ ☐ $1\frac{3}{5}$	e. $\frac{9}{5}$ ☐ $\frac{6}{6}$
f. $\frac{5}{12}$ ☐ $\frac{5}{11}$	g. $\frac{12}{8}$ ☐ $\frac{8}{12}$	h. $\frac{5}{5}$ ☐ $\frac{5}{7}$	i. $\frac{2}{3}$ ☐ $\frac{5}{10}$	j. $2\frac{4}{8}$ ☐ $2\frac{2}{7}$

> **Sometimes you can write an equivalent fraction, then compare.**
>
> **Example 3.** Compare $\frac{3}{5}$ and $\frac{7}{10}$. This time, we can write 3/5 as 6/10 since they are equivalent fractions. Then the problem changes to comparing $\frac{6}{10}$ and $\frac{7}{10}$, and clearly 7/10 is more.

7. Compare the fractions. Write an equivalent fraction for one of the fractions.
   Hint: To make an equivalent fraction, multiply both the top and bottom number in the fraction by some same number.

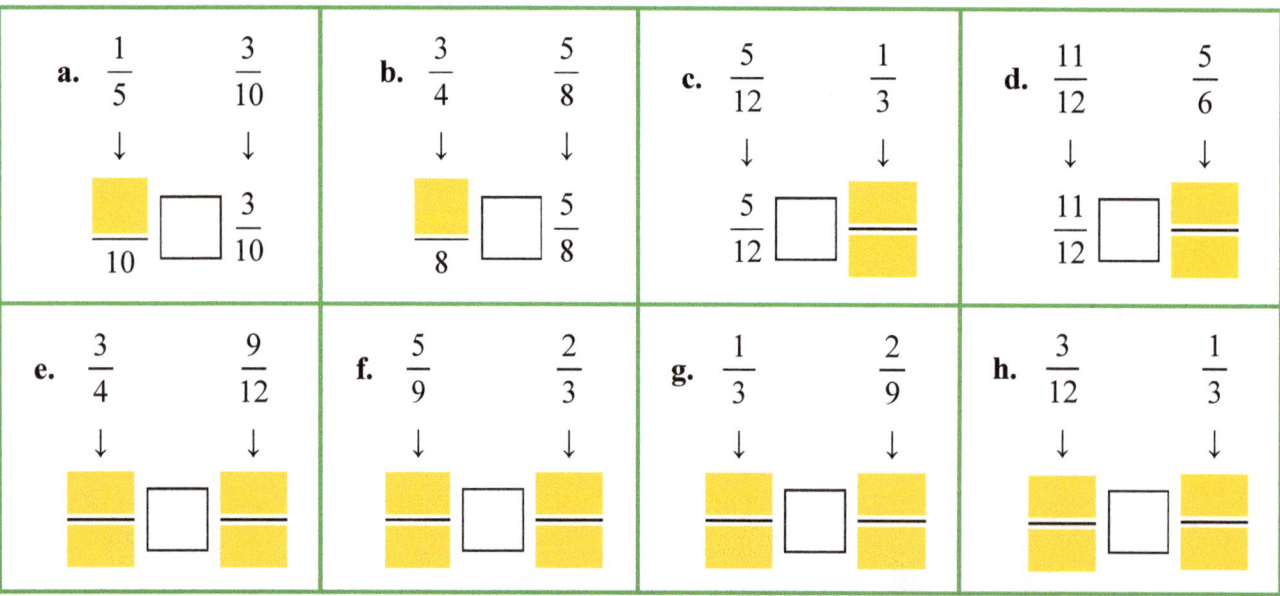

You cannot always base fraction comparisons on images.

Example 4. 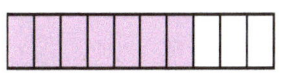 $\frac{7}{10}$

$\frac{5}{8}$

It *looks like* 7/10 is the same length as 5/8, but the fraction bars themselves are not the same size. So, we cannot say that 7/10 = 5/8!

When comparing fraction images, the wholes that the fractions are part of need to be equal in size.

8. Can you compare the fractions based on the images? If yes, write <, >, or = . If not, state that.

a. $\frac{7}{12}$ $\frac{6}{9}$

b. $\frac{3}{9}$ $\frac{2}{6}$

c. $\frac{7}{10}$ $\frac{5}{8}$

d. $\frac{5}{6}$ $\frac{7}{9}$

e. $\frac{7}{8}$ $\frac{5}{6}$

f. $\frac{2}{3}$ $\frac{7}{9}$

9. Arrange these fractions in order from the smallest to the greatest. Use the number lines to help.

$\frac{5}{8}, \frac{1}{3}, \frac{2}{5}, \frac{3}{8}, \frac{2}{3}$

10. Draw a picture to show that 1/3 < 1/2. You can use lines, bars, circles, or other shapes.

11. One number line here is divided into fifths. Divide the other into sixths, and then use the number lines to show that 5/6 > 3/5.

    Hint: First divide the number line into two halves. Then, divide each half into three parts.

12. Angie ate 3/8 of a pizza, and Joe ate 1/4 of the same pizza.
    Who ate more pizza?

    How much more pizza?

13. Bob pays 21/100 of his paycheck in taxes, and Chloe pays 3/10 of hers in taxes.
    Who pays a bigger part of his/her paycheck in taxes?

14. The store is having a sale! Which is a bigger discount:
    if a bike is discounted by 35/100 of its price,
    or if it is discounted by 4/10 of its price?

15. **a.** Emily drew these pictures, trying to show
    that 3/9 is more than 3/8. What is wrong?

    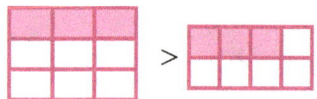

    b. Draw a picture showing that,
    actually, the opposite is true.

16. Write these fractions and mixed numbers in order, from the smallest to the greatest.

a. $\frac{3}{7}, \frac{3}{5}, 1\frac{1}{7}$	**b.** $1\frac{1}{4}, \frac{3}{8}, \frac{3}{6}$	**c.** $\frac{2}{3}, \frac{4}{9}, \frac{6}{5}$

Puzzle Corner Mom baked two rectangular pizzas. One was twice as big as the other.
Bob ate 2/3 of the smaller pizza, and Dad ate 3/8 of the larger pizza.

Who ate more pizza?

Explain your reasoning.

Equivalent Fractions 1

If you eat half of a pizza, or 2/4 of a pizza, you have eaten the same amount. The two fractions are *equivalent*.

We can write an equal sign between them: $\frac{1}{2} = \frac{2}{4}$.

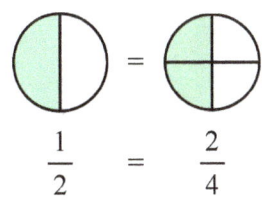

The dot for $\frac{3}{5}$ is in the same place on the number line as the dot for $\frac{6}{10}$. Again, the two fractions are *equivalent*. We can write $\frac{3}{5} = \frac{6}{10}$.

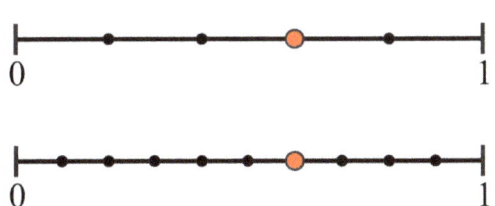

1. Write the equivalent fractions.

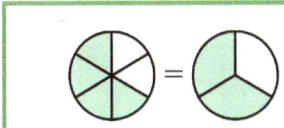

a. ——— = ———   b. ——— = ———

c. ——— = ———

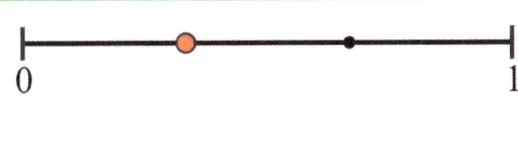

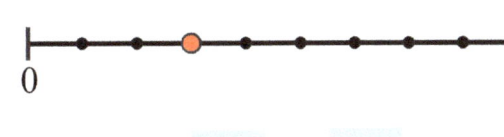

d. ——— = ———   e. ——— = ———

2. Shade the parts for the first fraction. Shade the same *amount* in the second picture. Write the second, equivalent fraction.

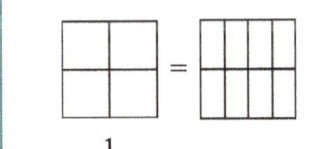

a. $\frac{1}{4} =$ b. $\frac{2}{4} =$ c. $\frac{6}{8} =$ d. $\frac{2}{3} =$

3. Draw an illustration to show the equivalence of the fractions. You can use any fraction model you feel works the best.

a. $\dfrac{3}{4} = \dfrac{6}{8}$	b. $\dfrac{1}{3} = \dfrac{2}{6}$

4. Write at least three fractions that are equivalent to 1/2. Also, use illustrations to show why they are equivalent.

5. Find all the pictures that show a fraction equivalent to 3/4.

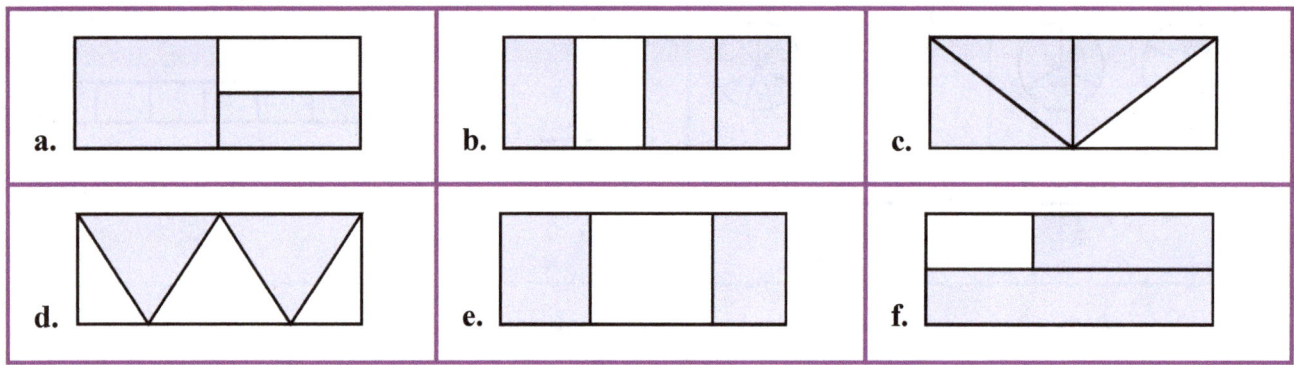

6. Are 3/3 and 4/4 equivalent fractions? Why or why not?

7. Shade a fraction that is equivalent to the given fraction.

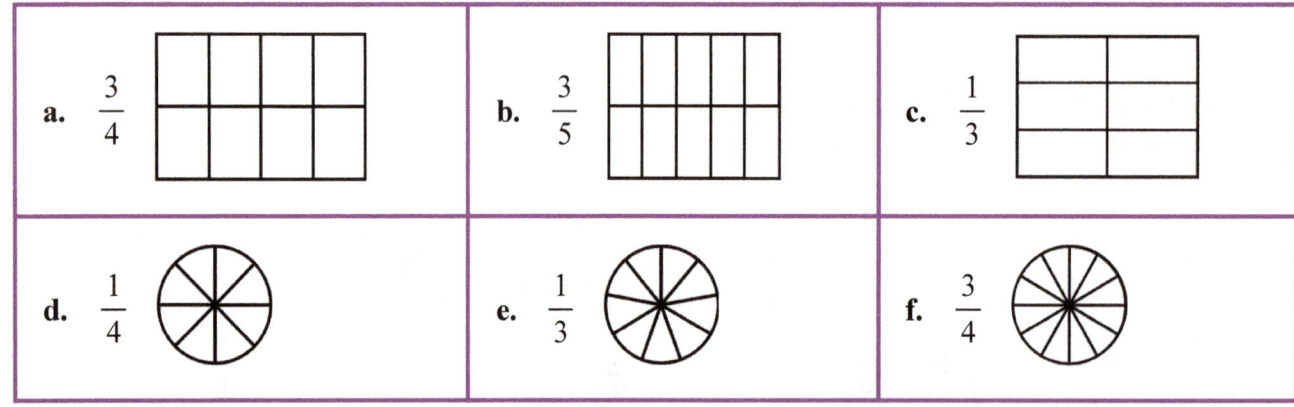

Equivalent Fractions 2

Besides writing one whole as $\frac{2}{2}$ or $\frac{6}{6}$, we can also write it as $\frac{1}{1}$.

The top "1" signifies how many pieces there are — one.
The bottom "1" signifies how many parts the whole is divided into — just one.

Similarly, 4 can be written as $\frac{4}{1}$. As a picture, this is 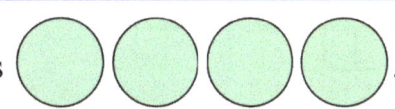.

When you think of the fraction line as a division symbol, this makes sense:
$\frac{4}{1}$ is the same as 4 ÷ 1 which equals 4.

1. Partition each number line into the given parts.

2. Write the number 2 as a fraction in at least three different ways.
   Explain or show why your fractions are equivalent.

3. Use the number lines above to write fractions that are equivalent to these.

a. $\frac{1}{1}$ = ___ = ___ = ___

b. $\frac{1}{4}$ = ___

c. $\frac{3}{2}$ = ___ = ___

4. Color and write many fractions that are equivalent to the first fraction.

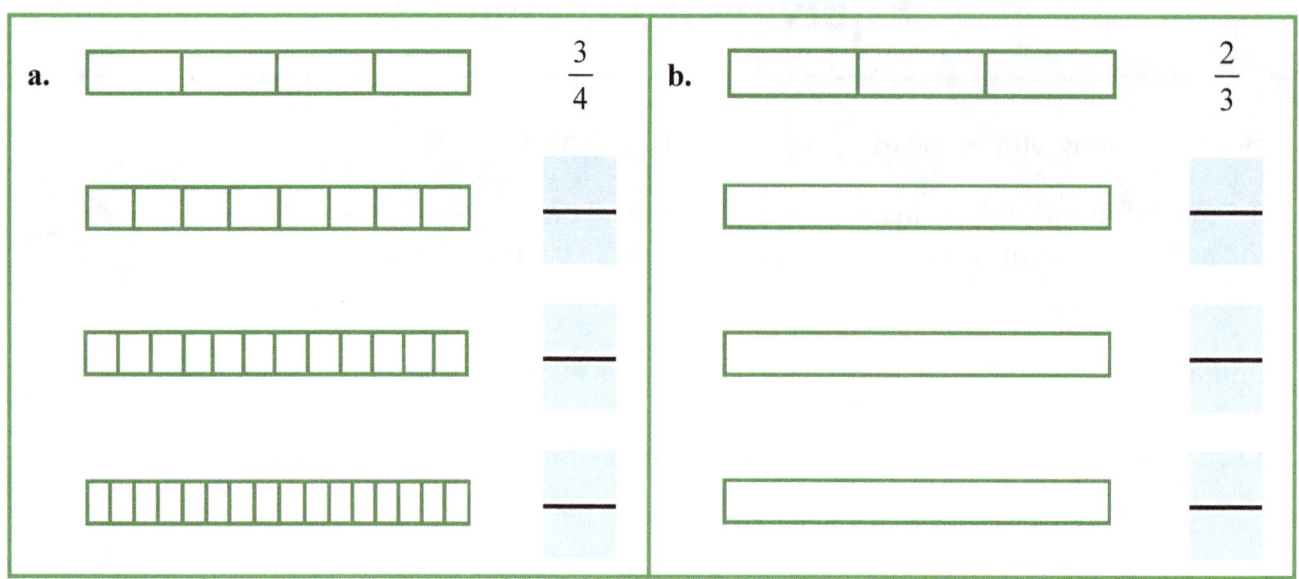

5. **a.** Find the fractions that are equal to 3: $\frac{6}{2}$ $\frac{2}{6}$ $\frac{4}{2}$ $\frac{6}{3}$ $\frac{3}{3}$ $\frac{9}{3}$ $\frac{3}{6}$ $\frac{3}{1}$

   b. Explain why they are equal to 3.
   (You may use pictures or explain it without pictures.)

6. Amanda and Joe each ordered a small pizza to eat. Amanda's was cut into six pieces and she ate three pieces. Joe's was cut into eight pieces. If Joe eats the same amount as Amanda, how many pieces would he eat?

7. Each square is one whole. Match the equivalent fractions. (One illustration will not be matched.)

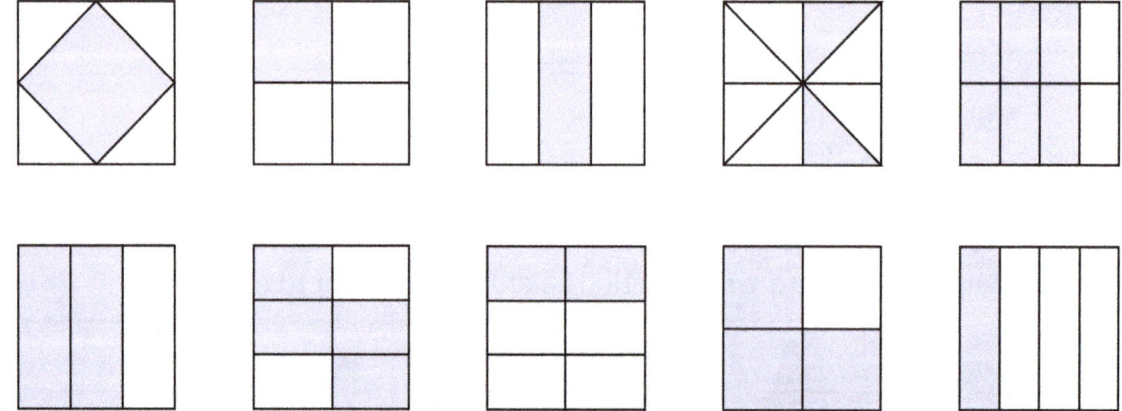

Equivalent Fractions 3
(This lesson is optional.)

1. The pie model on the right shows a certain fraction. Which of the following fractions are equivalent to it? $\frac{2}{3}$ $\frac{3}{6}$ $\frac{6}{2}$ $\frac{2}{4}$ $\frac{1}{3}$

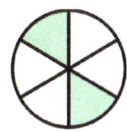

2. Some of these equations are not true. Correct the ones that are false, by changing one number in them.

a. $\frac{3}{3} = 3$	b. $\frac{3}{1} = 1$	c. $\frac{4}{2} = \frac{6}{3}$	d. $\frac{4}{10} = \frac{2}{5}$	e. $3 = \frac{12}{3}$

3. Draw illustrations to show different fractions that are equivalent to 1/3.

4. Liam says, "These fractions are equivalent." $\frac{2}{4} = \frac{1}{8}$

    He says that is so because 2 × 4 = 8. Explain to Liam why his thinking is wrong.

5. a. Half of the cherry pie is left. Show in the picture how three persons can share it equally.

   b. Your "cutting" shows certain two fractions to be equivalent. Which ones?

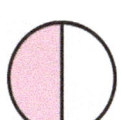

6. Show the equivalent fractions on the number lines.

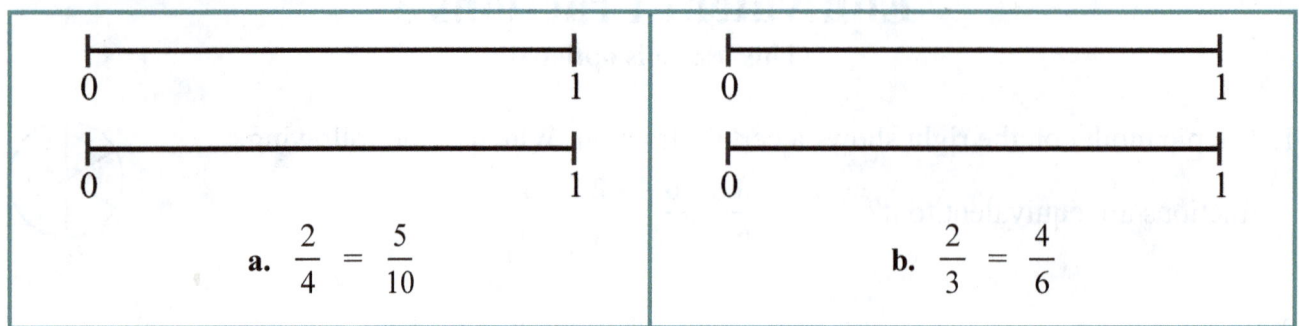

a. $\dfrac{2}{4} = \dfrac{5}{10}$

b. $\dfrac{2}{3} = \dfrac{4}{6}$

7. These pictures show four fruit bars. They illustrate that $4 = \dfrac{4}{1}$.

   a. Now "cut" each fruit bar into halves. Now the illustration shows that 4 is equal to another fraction. Which?

   b. Now "cut" each fruit bar in a different way, and write 4 as a fraction according to your illustration.

8. Match the equivalent fractions. Not all fractions will be matched.

$\dfrac{2}{4}$ $\dfrac{5}{10}$ $\dfrac{2}{6}$ $\dfrac{4}{2}$

$\dfrac{3}{4}$ $\dfrac{1}{3}$ $\dfrac{4}{8}$ $\dfrac{1}{5}$

$\dfrac{3}{1}$ $\dfrac{6}{8}$ $\dfrac{2}{1}$ $\dfrac{8}{4}$

$\dfrac{10}{2}$ $\dfrac{5}{1}$ $\dfrac{7}{10}$ $\dfrac{16}{8}$

Puzzle Corner   Which whole numbers are these?

a. 60 fourths   b. 104 eighths   c. 120 halves   d. 75 thirds

Equivalent Fractions 4

If you eat half of a pizza, or if you eat 4/8 of a pizza, you have eaten the same amount.

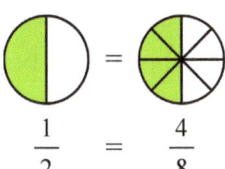

1/2 and 4/8 are **equivalent fractions**.

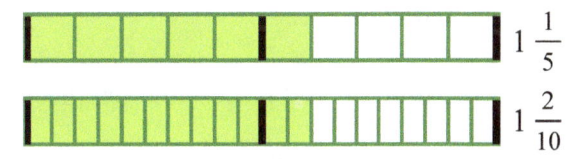

The two fraction strips show an equal amount. So, we can write an equal sign between the two mixed numbers: $1\frac{1}{5} = 1\frac{2}{10}$.

1. Color the first fraction. Shade the same amount of pie in the second picture. Write the second fraction.

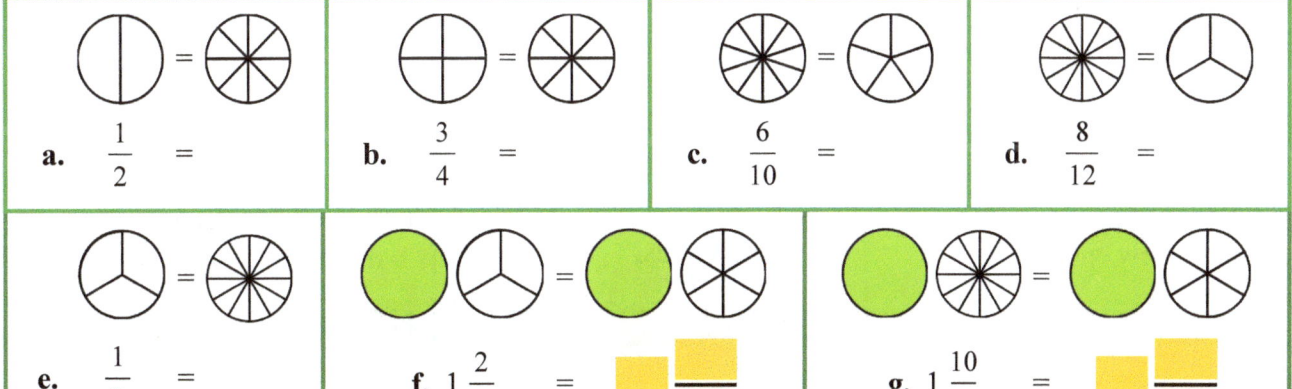

a. $\dfrac{1}{2} =$

b. $\dfrac{3}{4} =$

c. $\dfrac{6}{10} =$

d. $\dfrac{8}{12} =$

e. $\dfrac{1}{3} =$

f. $1\dfrac{2}{3} =$

g. $1\dfrac{10}{12} =$

2. Write the fractions that have thirds using sixths instead. You can shade parts in the pictures.

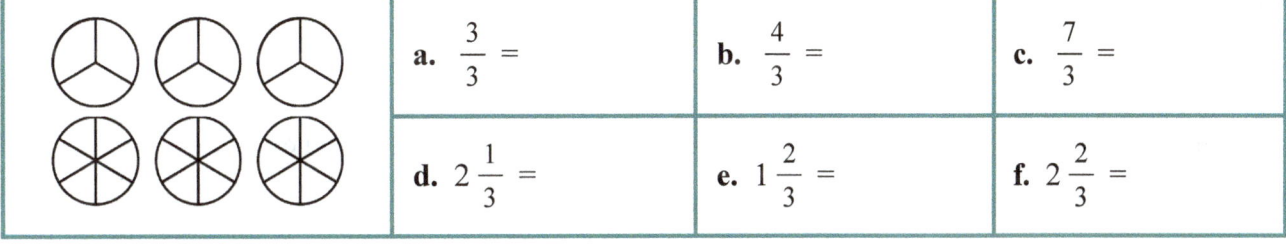

a. $\dfrac{3}{3} =$

b. $\dfrac{4}{3} =$

c. $\dfrac{7}{3} =$

d. $2\dfrac{1}{3} =$

e. $1\dfrac{2}{3} =$

f. $2\dfrac{2}{3} =$

3. Mark the equivalent fractions on the number lines.

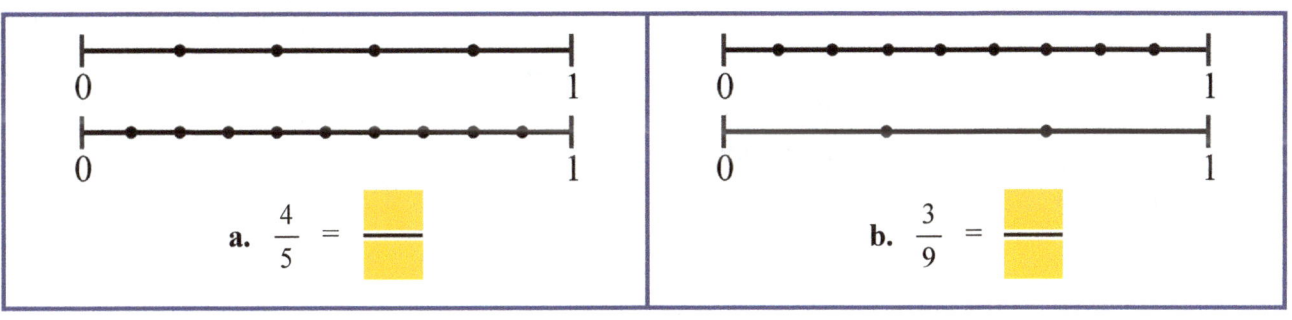

a. $\dfrac{4}{5} =$

b. $\dfrac{3}{9} =$

Example 1. The fraction strip illustrates $\frac{2}{5}$. If you split each piece (both the colored and white pieces) into *two* new pieces, what fraction do you get?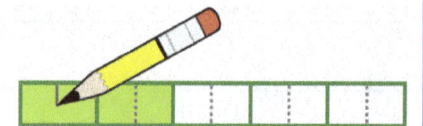

You get $\frac{4}{10}$: four colored pieces, and ten pieces total.

You have *two* times as many colored pieces, and *two* times as many total pieces as before.

4. Split both the colored and white pieces as instructed. Write the fraction after you change it.

a. Split all the pieces into two new ones. $\frac{1}{2} = $ ___	**b.** Split all the pieces into four new ones. $\frac{1}{2} = $ ___	**c.** Split all the pieces into three new ones. $\frac{1}{4} = $ ___
d. Split all the pieces into three new ones. $\frac{1}{3} = $ ___	**e.** Split all the pieces into two new ones. $\frac{5}{6} = $ ___	**f.** Split all the pieces into three new ones. $\frac{2}{5} = $ ___

Do you notice a *shortcut* for finding the second fraction?

g. Split all the pieces into four new ones. ___ = ___	**h.** Split all the pieces into two new ones. ___ = ___	**i.** Split all the pieces into three new ones. ___ = ___
If you found the shortcut, explain how it works in these problems:	Split all the pieces into three new ones. $\frac{1}{3} = $ ___	Split all the pieces into two new ones. $\frac{3}{5} = $ ___

Example 2. The fraction strip illustrates $\frac{1}{2}$. If we split each piece into *three* new pieces, we get $\frac{3}{6}$.

Now we have *three* times as many colored pieces, and *three* times as many pieces in total as we had before. Look at the right side of this box, to see how we can illustrate it this way →

We multiply both the top and bottom number in a fraction by 3. We get an equivalent fraction—it is the same amount, just cut into more pieces. *This does not mean we multiply the whole fraction by 3.*

$$\frac{1}{2} \overset{\times 3}{\underset{\times 3}{=}} \frac{3}{6}$$

5. Split the pieces. Fill in the missing parts.

a. This is $\frac{3}{4}$. Make it $\frac{9}{12}$.

Each piece is split into ____ new ones.

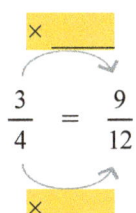

$$\frac{3}{4} = \frac{9}{12}$$

b. This is $\frac{1}{3}$. Make it $\frac{4}{12}$.

Each piece is split into ____ new ones.

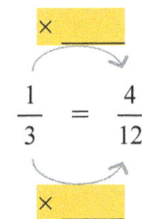

$$\frac{1}{3} = \frac{4}{12}$$

c. This is $\frac{1}{2}$. Make it $\frac{5}{10}$.

Each piece is split into ____ new ones.

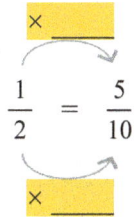

$$\frac{1}{2} = \frac{5}{10}$$

d. This is $\frac{1}{4}$. Make it $\frac{4}{16}$.

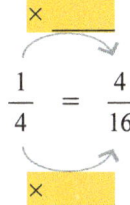

$$\frac{1}{4} = \frac{4}{16}$$

e. This is $\frac{2}{3}$. Make it $\frac{6}{9}$.

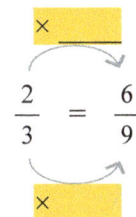

$$\frac{2}{3} = \frac{6}{9}$$

f. This is $\frac{2}{3}$. Make it $\frac{8}{12}$.

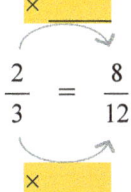

$$\frac{2}{3} = \frac{8}{12}$$

g.

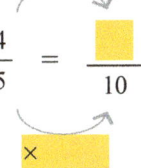

$$\frac{4}{5} = \frac{\boxed{}}{10}$$

h.

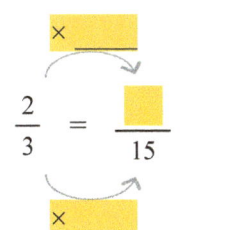

$$\frac{2}{3} = \frac{\boxed{}}{15}$$

i.

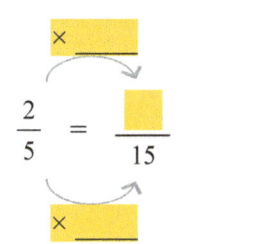

$$\frac{2}{5} = \frac{\boxed{}}{15}$$

6. Write the equivalent fraction. Use multiplication.

a. Split all the pieces into three new ones.	b. Split all the pieces into five new ones.	c. Split all the pieces into four new ones.	d. Split all the pieces into ten new ones.
$\frac{5}{6} = \frac{\ }{\ }$	$\frac{3}{4} = \frac{\ }{\ }$	$\frac{2}{5} = \frac{\ }{\ }$	$\frac{9}{10} = \frac{\ }{\ }$

7. Figure out how many new pieces the existing pieces were split into. Fill in the missing parts.

a. Pieces were split into ____ new ones. $\frac{1}{2} = \frac{\ }{6}$	b. Pieces were split into ____ new ones. $\frac{3}{10} = \frac{30}{\ }$	c. Pieces were split into ____ new ones. $\frac{2}{5} = \frac{\ }{30}$	d. Pieces were split into ____ new ones. $\frac{7}{8} = \frac{35}{\ }$
e. $\frac{2}{3} = \frac{\ }{6}$	f. $\frac{3}{5} = \frac{9}{\ }$	g. $\frac{5}{6} = \frac{\ }{12}$	h. $\frac{1}{3} = \frac{\ }{9}$

8. Write the fractions that have tenths with hundredths instead.

a. $\frac{1}{10} = \frac{\ }{100}$	b. $\frac{3}{10} =$	c. $\frac{6}{10} =$	d. $\frac{4}{10} =$	e. $\frac{13}{10} =$

9. Connect the equivalent fractions with a line.

a.
$\frac{2}{3}$ $\frac{1}{3}$
$\frac{1}{4}$ $\frac{1}{2}$
$\frac{5}{10}$ $\frac{2}{8}$
$\frac{2}{6}$ $\frac{6}{9}$

b.
$\frac{1}{2}$ $\frac{2}{10}$
$\frac{3}{4}$ $\frac{1}{3}$
$\frac{1}{5}$ $\frac{6}{12}$
$\frac{4}{12}$ $\frac{9}{12}$

c.
$\frac{3}{6}$ $\frac{3}{12}$
$\frac{1}{4}$ $\frac{1}{2}$
$\frac{1}{3}$ $\frac{8}{12}$
$\frac{2}{3}$ $\frac{4}{12}$

10. Write chains of equivalent fractions!

a. $\frac{1}{2} = \frac{\ }{4} = \frac{\ }{6} = \frac{\ }{8} = \frac{\ }{\ } = \frac{\ }{\ } = \frac{\ }{\ }$

b. $\frac{1}{3} = \frac{\ }{6} = \frac{\ }{9} = \frac{\ }{12} = \frac{\ }{\ }$

We can use equivalent fractions to add fractions that have different denominators.

Example 3. Add $\frac{2}{10} + \frac{17}{100}$. First, write 2/10 as 20/100 (an equivalent fraction). Then you can add, because the fractions now have the same denominator: $\frac{20}{100} + \frac{17}{100} = \frac{37}{100}$.

11. Add.

a. $\frac{1}{10} + \frac{8}{100}$ $\downarrow \quad \downarrow$ $\frac{\square}{100} + \frac{8}{100} =$	**b.** $\frac{7}{10} + \frac{3}{100}$ $\downarrow \quad \downarrow$ $\frac{\square}{100} + \frac{\square}{100} =$	**c.** $\frac{45}{100} + \frac{3}{10}$
d. $\frac{9}{10} + \frac{9}{100}$	**e.** $\frac{7}{10} + \frac{23}{100}$	**f.** $\frac{24}{100} + \frac{9}{10}$
g. $\frac{7}{100} + 1\frac{4}{10}$	**h.** $2\frac{28}{100} + 1\frac{5}{10}$	**i.** $\frac{6}{10} + \frac{35}{100} + \frac{7}{100}$

12. Draw a picture showing that 1/3 and 4/12 are equivalent fractions.

Puzzle Corner Add. *This is challenging. Hint: You cannot simply add the top numbers and the bottom numbers. Use equivalent fractions.*

a. $\frac{3}{4} + \frac{1}{2}$	**b.** $\frac{1}{5} + \frac{3}{10}$	**c.** $\frac{2}{3} + \frac{2}{9}$

Adding Fractions

It is easy to add fractions that have the same kinds of parts. Study the examples.

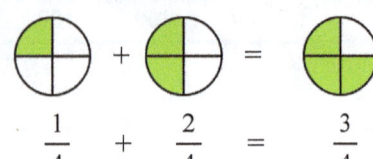

Example 1. Here, think of the pie pieces or slices. One fourth means one piece, and two fourths means two pieces. In total we have three pieces, and they all are fourths. So, the answer is 3/4.

$$\frac{1}{4} + \frac{2}{4} = \frac{3}{4}$$

Example 2. In this picture we have shaded (added) seven slices and then another six slices. All the slices are eighth parts so we can simply count how many eighths we get: 13 eighths or 13/8.

$$\frac{7}{8} + \frac{6}{8} = \frac{13}{8} = 1\frac{5}{8}$$

That makes more than one whole pie, so the answer is given as a mixed number: 1 5/8.

1. Solve. You can shade parts to help you. Give your answer as a mixed number when possible.

a. $\dfrac{1}{6} + \dfrac{3}{6} =$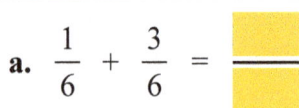

b. $\dfrac{2}{8} + \dfrac{5}{8} =$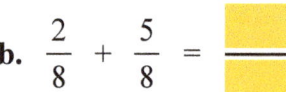

c. $\dfrac{7}{8} + \dfrac{7}{8} =$

d. $\dfrac{7}{10} + \dfrac{5}{10} =$

This is a fraction strip:

The shaded parts illustrate the addition $\dfrac{4}{5} + \dfrac{4}{5} = \dfrac{8}{5} = 1\dfrac{3}{5}$.

2. Add. Shade parts with different colors. Give your answer as a mixed number.

a. $\dfrac{3}{5} + \dfrac{4}{5} =$

b. $1\dfrac{2}{5} + \dfrac{4}{5} =$

c. $\dfrac{13}{10} + \dfrac{6}{10} =$

d. $1\dfrac{3}{8} + \dfrac{6}{8} =$

Adding like fractions (fractions that have the same kinds of pieces) is very easy. Think of "slices" or "pie pieces", add the actual <u>number of slices</u>, and lastly check what *kind* of slices they were.

Example 1. $\frac{6}{11} + \frac{4}{11} = ?$ Simply add 6 + 4 = 10 to find the total <u>number of slices.</u>

Since the slices are all eleventh parts, the answer is also: $\frac{6}{11} + \frac{4}{11} = \frac{10}{11}$.

Example 2. $\frac{3}{5} + \frac{4}{5} + \frac{1}{5} = \frac{8}{5} = 1\frac{3}{5}$

Here the answer, 8 fifths, is more than one whole, so we give our final answer as a mixed number.

3. Add the fractions. Give your final answer as a whole number or mixed number if possible.

a. $\frac{1}{6} + \frac{1}{6} =$	b. $\frac{1}{4} + \frac{3}{4} =$	c. $\frac{2}{8} + \frac{1}{8} + \frac{4}{8} =$
d. $\frac{1}{5} + \frac{2}{5} + \frac{4}{5} =$		e. $\frac{2}{3} + \frac{2}{3} + \frac{2}{3} =$
f. $\frac{11}{10} + \frac{7}{10} =$		g. $\frac{3}{4} + \frac{6}{4} + \frac{1}{4} =$

4. The children divided a chocolate bar into 12 pieces. Then, Mark ate 3/12 of it, Amy ate 2/12 of it, and Sam ate 4/12 of it.

   What fraction of the chocolate bar did the children eat?

   What fraction of the chocolate bar is left?

5. Add. Use the fraction strips if you need help.

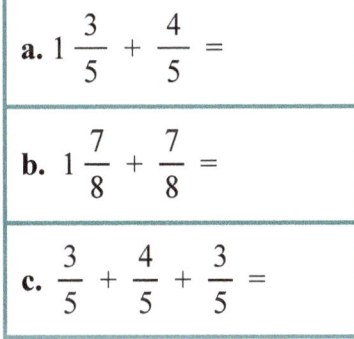

a. $1\frac{3}{5} + \frac{4}{5} =$

b. $1\frac{7}{8} + \frac{7}{8} =$

c. $\frac{3}{5} + \frac{4}{5} + \frac{3}{5} =$

6. Find what is missing from the additions. You can draw pie pictures on blank paper to help you.

a. $\frac{2}{12} +$ ___ $= \frac{11}{12}$	b. $\frac{5}{6} +$ ___ $= 1\frac{2}{6}$	c. $1\frac{5}{8} +$ ___ $= 2\frac{1}{8}$

Adding Mixed Numbers

Example 1.

Add the whole numbers and the fractions separately.

Since $\frac{10}{8}$ is $\frac{8}{8} + \frac{2}{8}$, or 1 and $\frac{2}{8}$,

the final answer becomes 7 2/8.

$$2\frac{3}{8} + 4\frac{7}{8} = 6\frac{10}{8}$$
$$= 6\frac{8}{8} + \frac{2}{8}$$
$$= 7\frac{2}{8}$$

1. Add the mixed numbers. You can shade parts to help.

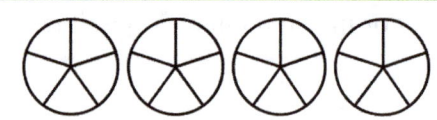

a. $1\frac{3}{5} + 2\frac{2}{5} =$

b. $1\frac{3}{7} + 2\frac{6}{7} =$

c. $1\frac{3}{8} + 1\frac{6}{8} =$

d. $\frac{8}{9} + 1\frac{5}{9} =$

2. Add the mixed numbers.

a. $1\frac{2}{5} + 3\frac{3}{5} =$

b. $4\frac{2}{6} + 2\frac{5}{6} =$

c. $5\frac{2}{4} + 7\frac{3}{4} =$

d. $1\frac{3}{8} + 8\frac{7}{8} =$

e. $7\frac{1}{6} + 20\frac{1}{6} =$

f. $8\frac{9}{10} + 3\frac{3}{10} =$

3. Pretend you are the "teacher" and find the errors in the these students' work! Then fix them.

a. Emma:

$1\frac{5}{7} = \frac{2}{7} + 1\frac{2}{7} + \frac{2}{7}$

$1\frac{5}{7} = \frac{10}{7} + \frac{2}{7}$

b. Peter: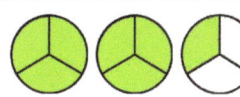

$2\frac{1}{3} = \frac{2}{3} + \frac{2}{3} + \frac{1}{3} + \frac{1}{3}$

$2\frac{1}{3} = \frac{5}{3} + \frac{3}{3}$

4. Write each mixed number as an addition in different ways.

a.

 =

 =

 =

 =

 =

 =

b.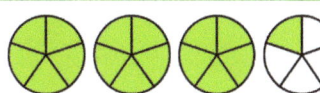

 =

 =

 =

 =

5. Solve.

a. A recipe calls for 1 ½ cups of wheat flour, ½ cup of rye flour, and ½ cup of oat flour. How much flour in total does the recipe use?

b. A movie lasted 1 ¾ hours and a meal afterwards took 1 ¼ hours. How much time in total did these take?

c. Jack drank 1 ¼ cups of water and ¾ cups of juice. How much liquid did he drink in total?

6. The sides of a rectangle are 2 ¼ inches and 3 ¼ inches long. Draw a sketch. What is its perimeter?

7. Each side of a triangle is 2 3/8 inches. What is its perimeter?

8. Double the cake recipe.

A birthday cake

4 eggs
3/4 cup sugar
1 1/4 cup flour
1 1/2 tsp baking powder
1 cup whipped cream
sliced fruit

A birthday cake

_____ eggs
_____ cup sugar
_____ cup flour
_____ tsp baking powder
_____ cup whipped cream
sliced fruit

Puzzle Corner 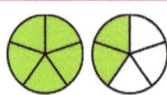   The picture shows the first addend. Draw more pies to figure out how much is missing from the additions.

a. $1\frac{2}{5} + \underline{} = 2\frac{3}{5}$

b. $1\frac{2}{4} + \underline{} = 3\frac{1}{4}$

c. $2\frac{5}{6} + \underline{} = 4\frac{2}{6}$

Subtracting Fractions and Mixed Numbers

Here are five sixths. If you take away two of them, how many sixths are left? Three sixths, of course!

Example 1. $\dfrac{5}{6} - \dfrac{2}{6} = \dfrac{3}{6}$

In this example, we can simply subtract the whole numbers and the fractions separately: 6 − 1 = 5, and $\dfrac{7}{8} - \dfrac{2}{8} = \dfrac{5}{8}$. It works, because from 7 eighths we *can* take away 2 eighths. If it was the other way around, we would need a different approach.

Example 2. $6\dfrac{7}{8} - 1\dfrac{2}{8} = 5\dfrac{5}{8}$

1. Subtract. You can cross out parts from the images to help you.

a. $\dfrac{9}{10} - \dfrac{1}{10} =$	**b.** $\dfrac{11}{12} - \dfrac{7}{12} =$	**c.** $2\dfrac{4}{6} - \dfrac{2}{6} =$ 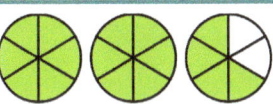
d. $2\dfrac{5}{9} - 1\dfrac{3}{9} =$ 	**e.** $\dfrac{9}{4} - \dfrac{3}{4} =$	**f.** $2\dfrac{7}{8} - \dfrac{3}{8} =$
g. $5\dfrac{9}{12} - 2\dfrac{5}{12} =$	**h.** $\dfrac{7}{10} - \dfrac{5}{10} =$	**i.** $10\dfrac{7}{12} - 7\dfrac{3}{12} =$ **j.** $1 - \dfrac{7}{8} =$

Example 3. 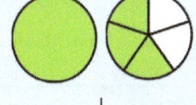 $1\dfrac{3}{5} - \dfrac{4}{5}$

↓   ↓

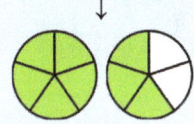 $\dfrac{8}{5} - \dfrac{4}{5} = \dfrac{4}{5}$

Here it helps to first **change the mixed number 1 3/5 into the fraction 8/5**. It is like cutting the whole pie into fifths. Now it is easy to subtract 4 fifths.

2. Subtract. To help you, you can "cut" the whole pie into pieces first.

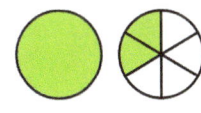	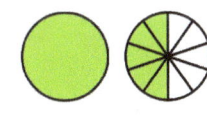	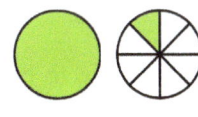	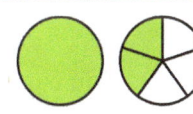
a. $1\dfrac{2}{6} - \dfrac{5}{6}$	**b.** $1\dfrac{5}{10} - \dfrac{9}{10}$	**c.** $1\dfrac{1}{8} - \dfrac{6}{8}$	**d.** $1\dfrac{2}{5} - \dfrac{4}{5}$

Example 4.

$3\frac{1}{5} - 1\frac{4}{5}$

↓

$2\frac{6}{5} - 1\frac{4}{5} = 1\frac{2}{5}$

First **change one of the whole pies into a fraction**. Then it is easy to subtract (whole numbers and fractional parts separately).

3. Subtract using the above method.

a. $3\frac{2}{10} - \frac{6}{10}$ ↓ ↓ $2\frac{12}{10} - \frac{6}{10} =$	b. $2\frac{1}{7} - \frac{5}{7}$	c. $5\frac{3}{9} - 2\frac{7}{9}$	d. $7\frac{2}{5} - 4\frac{4}{5}$

4. Subtract from whole numbers. To help you, you can first write one of the whole pies as a fraction.

a. $7 - \frac{3}{4}$ ↓ ↓ $6\frac{4}{4} - \frac{3}{4} =$	b. $6 - \frac{8}{9} =$	c. $5 - 1\frac{1}{2} =$	d. $10 - 2\frac{3}{5} =$

5. Mom made three whole pies for Thanksgiving, and divided each pie into twelfths. The children ate three slices before dinner. How much pie is left? Give your answer as a mixed number.

6. Subtract. You can use the fraction strip to help you.

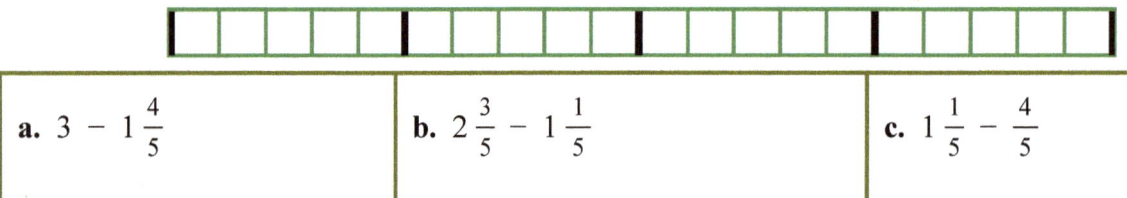

a. $3 - 1\frac{4}{5}$	b. $2\frac{3}{5} - 1\frac{1}{5}$	c. $1\frac{1}{5} - \frac{4}{5}$

Here I will show you two more methods for subtracting mixed numbers. It will be up to you what method you will use, whether the one explained above (method 1), or one of the two shown here, in the rest of the exercises of this lesson. We will use $3\frac{1}{5} - 1\frac{4}{5}$ as our example problem.

Method 2: subtract part by part.

First subtract 1 whole: $3\frac{1}{5} - 1$ leaves $2\frac{1}{5}$.

Then subtract 1 fifth: $2\frac{1}{5} - \frac{1}{5}$ leaves 2.

Lastly, subtract 3 fifths: $2 - \frac{3}{5} = 1\frac{2}{5}$.

Method 3: use fractions.

We write 3 1/5 and 1 4/5 as fractions, then subtract. Lastly, we write the answer as a mixed number:

$$3\frac{1}{5} - 1\frac{4}{5}$$
$$\downarrow \quad \downarrow$$
$$\frac{16}{5} - \frac{9}{5} = \frac{7}{5} = 1\frac{2}{5}$$

7. Subtract.

a. $2\frac{1}{4} - \frac{3}{4} =$	**b.** $3\frac{1}{8} - \frac{3}{8} =$	**c.** $6\frac{5}{6} - 3\frac{3}{6} =$
d. $2\frac{2}{5} - 1\frac{3}{5} =$	**e.** $7\frac{1}{5} - 2\frac{3}{5} =$	**f.** $6\frac{1}{3} - 3\frac{2}{3} =$

8. The perimeter of a rectangle is 5 inches. If one side measures 1 ½ inches, how long is the other side?

9. Subtract. First write the fractions that have tenths using hundredths instead.

a. $\dfrac{6}{10} - \dfrac{15}{100}$ ↓ ↓ $\dfrac{}{100} - \dfrac{15}{100} =$	b. $\dfrac{7}{10} - \dfrac{38}{100}$	c. $\dfrac{54}{100} - \dfrac{2}{10}$

10. Mary had 6 yards of material. She cut off a piece that was 2 2/3 yards long. How long is the part that is left, in yards?

    Challenge: How long are the two pieces in *feet*?

11. Three pizzas were ordered for a get-together. Each was divided into 12 pieces. Edward ate three pieces, Abigail ate two pieces, Mom and Dad together ate a whole pizza, and Jack and John ate four pieces each. How much of a pizza was left?

12. **a.** A bag of flour contains 6 cups of flour. Mom used 2 2/3 cups for a bread recipe. How much flour is left after she makes one batch of bread?

    b. How many batches more can she make with the remaining flour?

 The perimeter of a certain rectangle is 6 ½ inches. If one side measures 1 ¾ inches, how long is the other side?

Multiplying Fractions by Whole Numbers

You already know that $\frac{7}{8}$ is seven copies of $\frac{1}{8}$.

From this we can write a multiplication: $\frac{7}{8} = 7 \times \frac{1}{8}$.

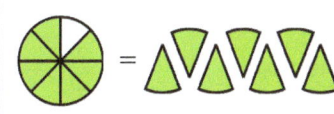

Similarly, $1\frac{1}{4}$ is $\frac{5}{4}$ (as a fraction), so it is five copies of $\frac{1}{4}$.

From this we can write a multiplication: $1\frac{1}{4} = 5 \times \frac{1}{4}$.

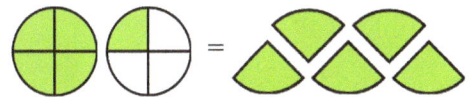

1. Fill in.

a. $\frac{3}{7} = 3 \times \underline{}$

b. $\frac{6}{9} = \boxed{} \times \underline{}$

c. $4 \times \frac{1}{5} = \underline{}$

d. $7 \times \frac{1}{10} = \underline{}$

2. Fill in.

a. $\frac{8}{7} = 8 \times \underline{}$

b. $1\frac{3}{5} = \frac{\boxed{}}{5} = 8 \times \underline{}$

c. $1\frac{2}{3} = \frac{\boxed{}}{3} = \boxed{} \times \underline{}$

d. $10 \times \frac{1}{6} = \underline{} = \boxed{} \underline{}$

e. $7 \times \frac{1}{4} = \underline{} = \boxed{} \underline{}$

f. $9 \times \frac{1}{3} = \underline{} =$

3. **a.** Mary is preparing a dinner for 10 people. She needs to buy 1/3 lb of chicken per person. How many pounds of chicken will she need to buy (at least)?

   b. Between what two whole numbers does your answer lie?

   c. She will also prepare 1/2 quart of juice for each guest. How many quarts of juice will she need?

Example 1. Look at the picture.

It illustrates $3 \times \frac{3}{4}$ as three copies of $\frac{3}{4}$.

How many fourths are there in total?

There are nine fourths. So, $3 \times \frac{3}{4} = \frac{9}{4}$.

As a mixed number, this is 2 ¼.

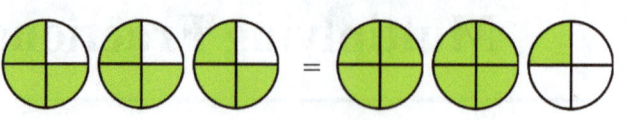

$3 \times \frac{3}{4} = \frac{9}{4} = 2\frac{1}{4}$

4. Multiply.

 a. $3 \times \frac{2}{4} = \frac{}{} = \square \frac{\square}{\square}$

 b. $4 \times \frac{2}{6} = \frac{}{} = \square \frac{\square}{\square}$

 c. $2 \times \frac{7}{8} = \frac{}{} = \square \frac{\square}{\square}$

5. Color repeatedly to solve the multiplications. *Give your answer as a mixed number.*

a. Color five times (copies of) $\frac{3}{8}$.

 $5 \times \frac{3}{8} =$

b. Color four times (copies of) $\frac{2}{5}$.

 $4 \times \frac{2}{5} =$

c. Color five times $\frac{7}{12}$.

 $5 \times \frac{7}{12} =$

d. Color five times $\frac{6}{10}$.

 $5 \times \frac{6}{10} =$

e. 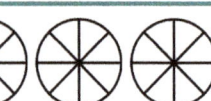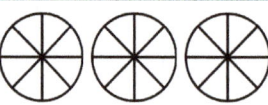 $9 \times \frac{5}{8} =$

Can you find a shortcut for these problems?
If you can, use it to solve the problems below. (*If not, don't worry about it.*)

f. $4 \times \frac{2}{3} =$

g. $3 \times \frac{4}{10} =$

h. $2 \times \frac{5}{6} =$

6. Fill in.

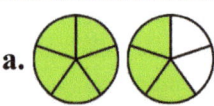 a. $\frac{8}{5} = 4 \times \frac{\square}{\square}$

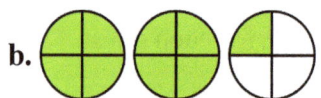 b. $\frac{9}{4} = 3 \times \frac{\square}{\square}$

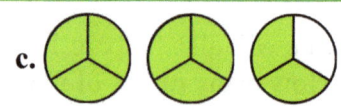 c. $2\frac{2}{3} = 2 \times \square \frac{\square}{\square}$

> **To multiply a whole number by a fraction, find the total number of "pieces".**
> This means you multiply the whole number and the top number of the fraction.
>
> **Example 2.** $6 \times \dfrac{2}{5}$ means 6×2 pieces, or 12 pieces. Each piece is a fifth. So, we get $\dfrac{12}{5}$.
> Lastly, change $\dfrac{12}{5}$ into a mixed number: it is $2\dfrac{2}{5}$.
>
> **Example 3.** Multiplication can be done in either order.
> So, $5 \times \dfrac{3}{8}$ is the same as $\dfrac{3}{8} \times 5$. They both equal $\dfrac{5 \times 3}{8} = \dfrac{15}{8}$, which is $1\dfrac{7}{8}$.

7. Solve. Give your answer as a mixed number if possible.

a. $5 \times \dfrac{1}{4} =$	b. $3 \times \dfrac{2}{3} =$	c. $4 \times \dfrac{2}{7} =$
d. $7 \times \dfrac{2}{10} =$	e. $\dfrac{3}{8} \times 6 =$	f. $7 \times \dfrac{2}{100} =$
g. $\dfrac{7}{10} \times 3 =$	h. $6 \times \dfrac{12}{100} =$	i. $\dfrac{11}{10} \times 3 =$
j. $\dfrac{5}{8} \times 5 =$	k. $2 \times \dfrac{4}{3} =$	l. $\dfrac{7}{4} \times 2 =$

8. The side of a square is 7/8 in. What is its perimeter?

9. Baby's toy blocks are 1 1/8 in. tall each.
    a. How tall is a stack of five of them?

    b. How about a stack of ten?

10. Jack prepares a pasta dish with meat for 8 people.
    He plans to buy 1/4 lb of meat per person.
    How much meat should he buy?

    He also needs 3/4 cups of dry pasta per person.
    How much pasta should he use?

Practicing with Fractions

1. Mark these fractions on the number line. *Hint: Think of equivalent fractions.*

$$\frac{5}{12}, \frac{11}{12}, \frac{1}{2}, \frac{1}{4}, \frac{2}{3}, \frac{5}{6}$$

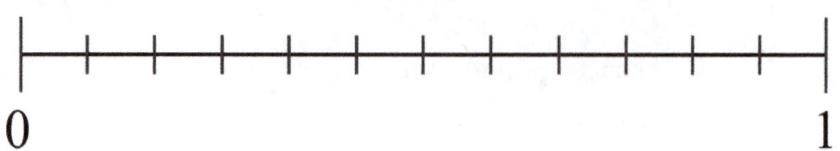

2. List the fractions in order from the smallest to the greatest. You can draw pie pictures to help.

a. $\frac{1}{2}, \frac{2}{3}, \frac{2}{6}$ ◯ ◯ ◯

b. $\frac{1}{4}, \frac{1}{8}, \frac{3}{8}$ ◯ ◯ ◯

c. $\frac{2}{5}, \frac{3}{5}, \frac{1}{2}$ ◯ ◯ ◯

d. $\frac{4}{5}, \frac{3}{8}, \frac{3}{4}$ ◯ ◯ ◯

3. Each day, Robert runs 3/5 of a mile. How far does he run in three days? Give your answer as a mixed number.

4. Make chains of equivalent fractions. You can choose the other two fractions but they need to be equivalent to the first.

a. ◯ = ◯ = ◯
   $\frac{1}{3}$ =

b. ◯ = ◯ = ◯
   $\frac{1}{4}$ =

c. ◯ = ◯ = ◯
   $\frac{3}{4}$ =

d. ◯ = ◯ = ◯
   $\frac{2}{5}$ =

5. Multiply, and find the missing numbers.

a. $2 \times \dfrac{2}{7} =$	b. $5 \times \dfrac{2}{3} =$	c. $3 \times \dfrac{3}{4} =$
d. ___ $\times \dfrac{4}{7} = \dfrac{8}{7}$	e. ___ $\times \dfrac{1}{2} = 3\dfrac{1}{2}$	f. ___ $\times \dfrac{3}{4} = 1\dfrac{2}{4}$

6. Add or subtract. Give your final answer as a whole number or as a mixed number.

a. $\dfrac{3}{9} + \dfrac{6}{9} =$	b. $\dfrac{5}{8} + \dfrac{6}{8} =$
c. $2\dfrac{2}{5} + 4\dfrac{4}{5} =$	d. $\dfrac{11}{12} - \dfrac{3}{12} =$
e. $1\dfrac{1}{10} - \dfrac{3}{10} =$	f. $5\dfrac{1}{4} - \dfrac{3}{4} =$

7. Multiplying by one-half is actually the same as _____ by 2. Fill in.

a.	b.	c.
$2 \times \dfrac{1}{2} =$	$7 \times \dfrac{1}{2} =$	$15 \times \dfrac{1}{2} =$
$3 \times \dfrac{1}{2} =$	$8 \times \dfrac{1}{2} =$	$20 \times \dfrac{1}{2} =$
$4 \times \dfrac{1}{2} =$	$9 \times \dfrac{1}{2} =$	$17 \times \dfrac{1}{2} =$
$5 \times \dfrac{1}{2} =$	$10 \times \dfrac{1}{2} =$	$21 \times \dfrac{1}{2} =$
$6 \times \dfrac{1}{2} =$	$11 \times \dfrac{1}{2} =$	$32 \times \dfrac{1}{2} =$

Puzzle Corner

Add *un*like fractions. First change one fraction by splitting its pieces further, then add.

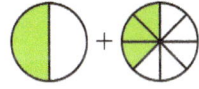

a. $\dfrac{1}{2} + \dfrac{3}{8} =$

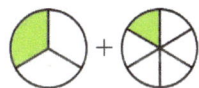

b.

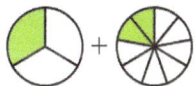

c.

Fractions Review

1. Divide each shape into parts and shade parts to illustrate each fraction.

   a. $\dfrac{5}{8}$ b. $\dfrac{7}{4}$ c. $\dfrac{4}{2}$

2. Write the fraction.

   a. ____   b. ____   c. ____   d. ____

3. Write the fraction shown by the big dot on the number line.

   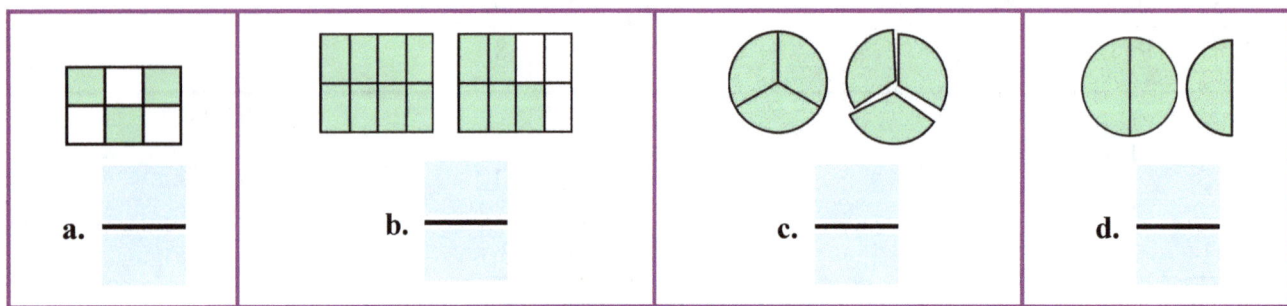

   a.   b.

4. Mark the fraction on the number line (with a dot).

   a.
   $\dfrac{2}{6}$

   b.
   $\dfrac{7}{8}$

5. Mark these fractions on the number line: $\dfrac{19}{6}, \dfrac{15}{6}, \dfrac{18}{6}, \dfrac{24}{6}, \dfrac{29}{6}$.

   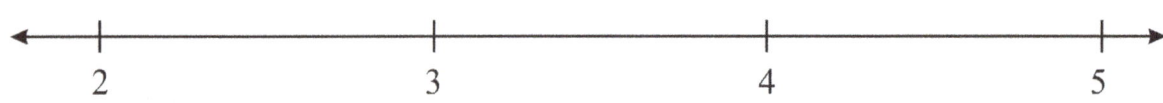

6. Between which two whole numbers is the fraction $\dfrac{35}{6}$?

7. Write the whole numbers as fractions.

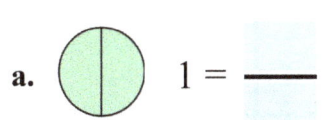 a. 1 = ―――
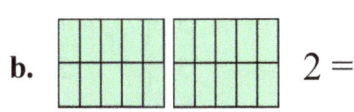 b. 2 = ―――
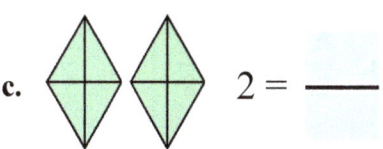 c. 2 = ―――

d. $4 = \dfrac{}{5}$ e. $7 = \dfrac{}{6}$ f. $4 = \dfrac{}{10}$ g. $6 = \dfrac{}{2}$ h. $5 = \dfrac{}{8}$

8. Write and shade the equivalent fractions.

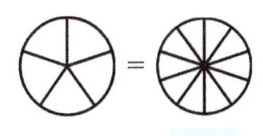 a. $\dfrac{2}{5} = $ ―――

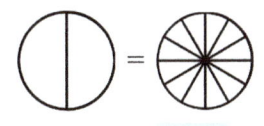 b. $\dfrac{1}{2} = $ ―――

c. $\dfrac{9}{12} = $ ―――

9. Write two other fractions that are equivalent to 1/3.

$$\dfrac{1}{3} = \underline{} = \underline{}$$

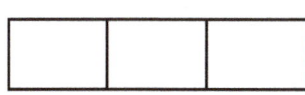
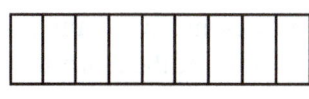

10. Show that $\dfrac{3}{4}$ and $\dfrac{6}{8}$ are equivalent fractions.

11. Compare the fractions, writing >, <, or = between them. If you cannot make a valid comparison, then cross the whole problem out.

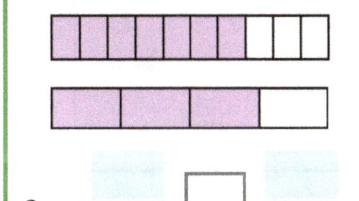
a. ――― ☐ ―――

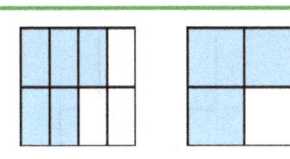
b. ――― ☐ ―――

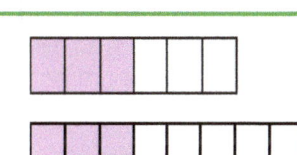
c. $\dfrac{3}{6}$ ☐ $\dfrac{3}{8}$

81

12. Compare the fractions. Write < , > or = between them.

| a. $\dfrac{6}{8}$ ☐ $\dfrac{7}{8}$ | b. $\dfrac{1}{5}$ ☐ $\dfrac{1}{8}$ | c. $\dfrac{3}{8}$ ☐ $\dfrac{3}{5}$ | d. $\dfrac{1}{2}$ ☐ $\dfrac{2}{4}$ | e. $\dfrac{24}{10}$ ☐ $\dfrac{15}{10}$ |

13. Explain how you can tell which is the greater fraction: $\dfrac{5}{2}$ or $\dfrac{5}{6}$?

    _____

    _____

14. Write these fractions in order from the smallest to the largest. The fraction bars can help.

    $\dfrac{1}{3}$ $\dfrac{3}{6}$ $\dfrac{2}{9}$ $\dfrac{4}{9}$

    ____ < ____ < ____ < ____

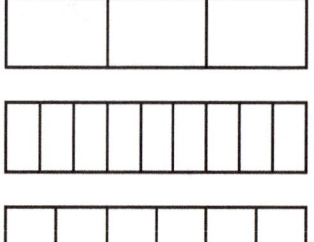
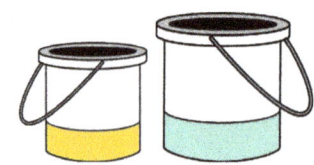

15. Jack has two paint cans, one bigger and one smaller. Both of them are 1/3 full.

    a. Which can has more paint in it?

    b. Does this mean that $\dfrac{1}{3} < \dfrac{1}{3}$?

       Why or why not?

a. Color 2/3 of this shape.   b. Color 3/4 of this shape.   c. Color 5/8 of this shape.

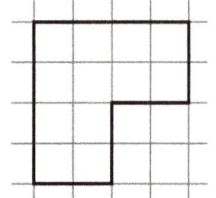 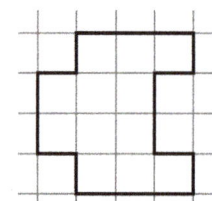 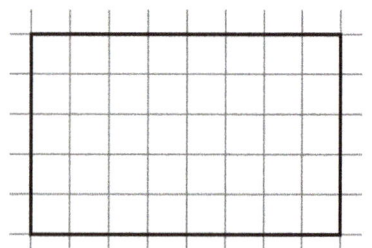

Review

1. Add or subtract. Give your final answer as a whole number or as a mixed number if possible.

a. $\frac{3}{5} + \frac{2}{5} =$	b. $4\frac{2}{8} + \frac{7}{8} =$	c. $2\frac{3}{4} + 4\frac{1}{4} =$
d. $\frac{9}{10} - \frac{7}{10} =$	e. $2\frac{1}{4} - \frac{3}{4} =$	f. $8\frac{9}{12} - 2\frac{2}{12} =$

2. Find the missing fractions.

a. $\frac{3}{10} + \underline{} = 1$	b. $3\frac{2}{5} + \underline{} = 4$	c. $\frac{4}{5} + \frac{3}{5} + \underline{} = 2\frac{1}{5}$	d. $7 - \underline{} = 6\frac{3}{8}$

3. Add.

a. $\frac{3}{10} + \frac{3}{100}$	b. $\frac{1}{10} + \frac{43}{100}$	c. $\frac{57}{100} + \frac{6}{10}$

4. Jane drank 1/4 liter of water from a full 2-liter pitcher. How much water is left in the pitcher?

5. Draw a picture showing that 2/5 = 4/10.

6. Write the equivalent fraction. Use multiplication.

a. Split all the pieces into two new ones. 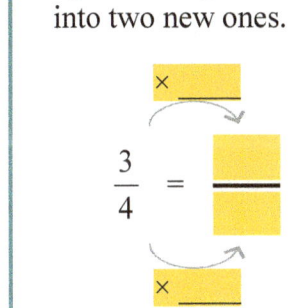	b. Split all the pieces into five new ones. 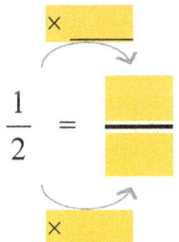	c. $\frac{2}{5} = \frac{4}{\boxed{}}$	d. $\frac{2}{3} = \frac{\boxed{}}{9}$
		e. $\frac{2}{3} = \frac{\boxed{}}{12}$	f. $\frac{3}{4} = \frac{12}{\boxed{}}$

7. Compare the fractions and mixed numbers.

a. $\dfrac{2}{9}$ ☐ $\dfrac{2}{10}$ b. $\dfrac{7}{10}$ ☐ $\dfrac{4}{5}$ c. $\dfrac{5}{10}$ ☐ $\dfrac{3}{6}$ d. $1\dfrac{1}{3}$ ☐ $1\dfrac{1}{5}$

e. $\dfrac{3}{4}$ ☐ $\dfrac{5}{8}$ f. $\dfrac{4}{9}$ ☐ $\dfrac{1}{3}$ g. $\dfrac{5}{12}$ ☐ $\dfrac{1}{3}$ h. $3\dfrac{4}{7}$ ☐ $3\dfrac{5}{6}$

8. Multiply.

a. $3 \times \dfrac{3}{10} =$	b. $3 \times \dfrac{2}{5} =$	c. $\dfrac{2}{10} \times 7 =$
d. $9 \times \dfrac{11}{100} =$	e. $4 \times \dfrac{5}{8} =$	f. $\dfrac{11}{12} \times 3 =$

9. Quadruple this recipe (*make it four times*).

Mexican Coffee

1 ½ cups strong gourmet coffee
¾ tsp cinnamon
4 tsp chocolate syrup
¼ tsp nutmeg
½ cup heavy cream
1 tbsp sugar

Mexican Coffee (4x)

_____ cups strong gourmet coffee
_____ tsp cinnamon
_____ tsp chocolate syrup
_____ tsp nutmeg
_____ cup heavy cream
_____ tbsp sugar

10. Remember division? Find the amounts.

a. $\dfrac{1}{5}$ of 200	b. $\dfrac{1}{6}$ of 48 cm	c. $\dfrac{2}{3}$ of 600 kg
$\dfrac{2}{5}$ of 200	$\dfrac{5}{6}$ of 48 cm	$\dfrac{5}{8}$ of $64

11. Mark got $20 for his birthday. The same day he spent 3/4 of it. Now how much does he have left?

12. Rose has read 3/8 of a 240-page book. How many pages are still left to read?

Answer Key

Halves and Quarters, pp. 9-12

Page 9

1.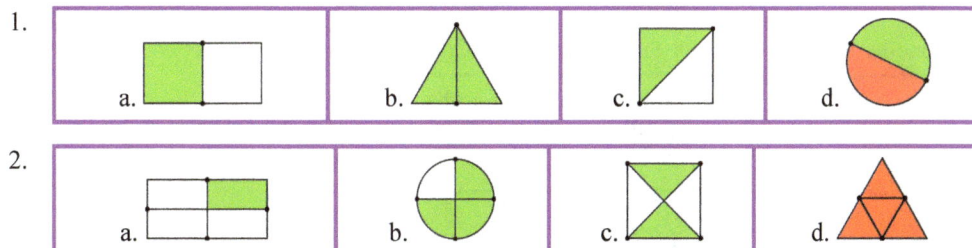

2.

Page 10

3.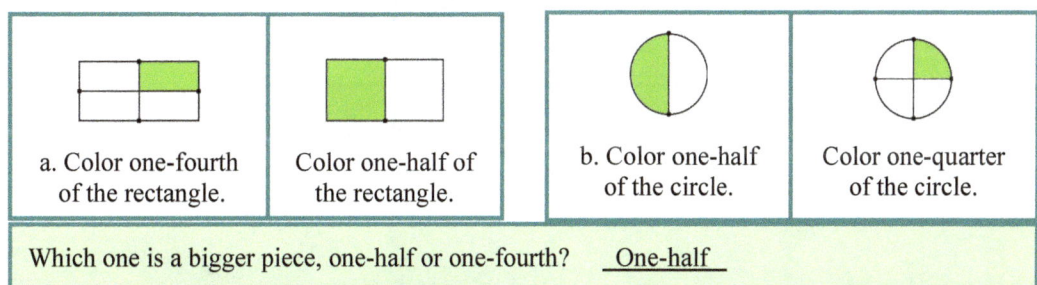

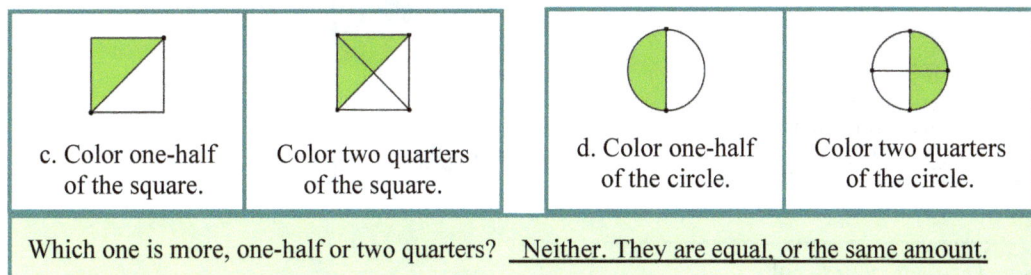

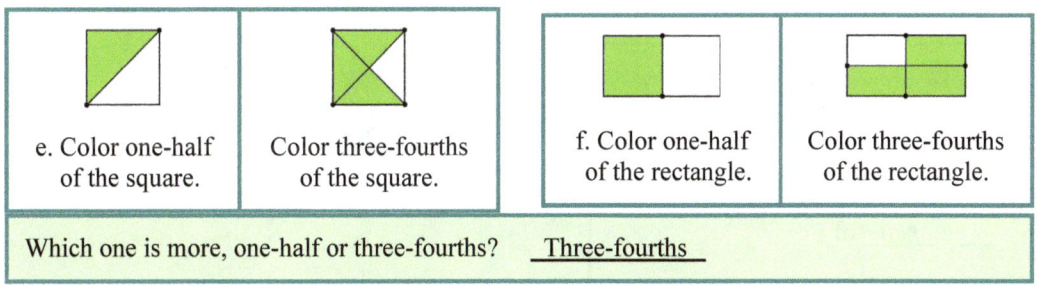

Page 11

4.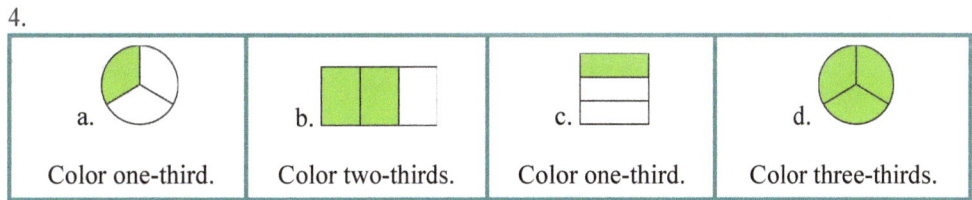

Halves and Quarters, cont.

Page 11

5.

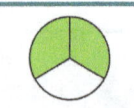 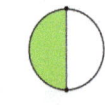

Color two-thirds.   Color one-half.

a. Which is more, two-thirds or one-half? two-thirds.

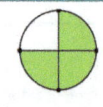 

Color three-fourths.   Color two-thirds.

b. Which is more, three-fourths or two-thirds? three-fourths.

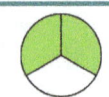 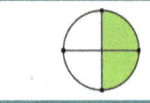

c. Which is more, two-thirds or two quarters? two-thirds.

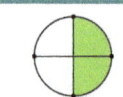 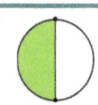

d. Which is more, two-fourths or one-half? Neither, they are equal.

6.

a. Which is more, one-half, or one-third? one-half.

b. Which is more, one-fourth, or one-third? one-third.

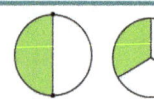

Page 12

7.

   a.

The whole pie is 3 thirds .

b.

The whole pie is 2 halves .

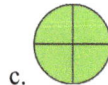   c.

The whole pie is 4 quarters/fourths.

8.

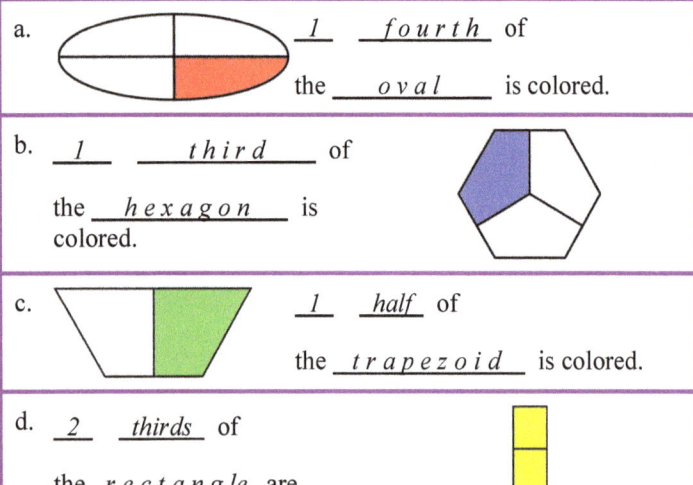

a. 1 fourth of the oval is colored.

b. 1 third of the hexagon is colored.

c. 1 half of the trapezoid is colored.

d. 2 thirds of the rectangle are colored.

Halves and Quarters, cont.

Page 12

8. continued.

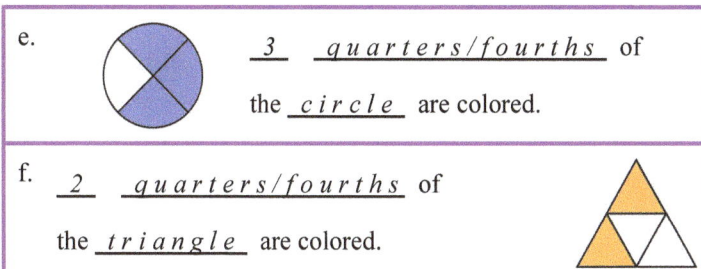

e. __3__ _quarters/fourths_ of the _circle_ are colored.

f. __2__ _quarters/fourths_ of the _triangle_ are colored.

Some Fractions, pp. 13-15

Page 13

1. a. b. c. d. e. f. g. h.

Page 14

2.
a. 4 little squares in one-half; 8 little squares in the whole.

b. 6 little squares in one-half; 12 little squares in the whole.

c. 1 little square in one-fourth; 4 little squares in the whole.

d. 2 little squares in one-fourth; 8 little squares in the whole.

e. 9 little squares in three-fourths; 12 little squares in the whole.

f. 8 little squares in two-thirds; 12 little squares in the whole.

Some Fractions, cont.

Page 15

3.

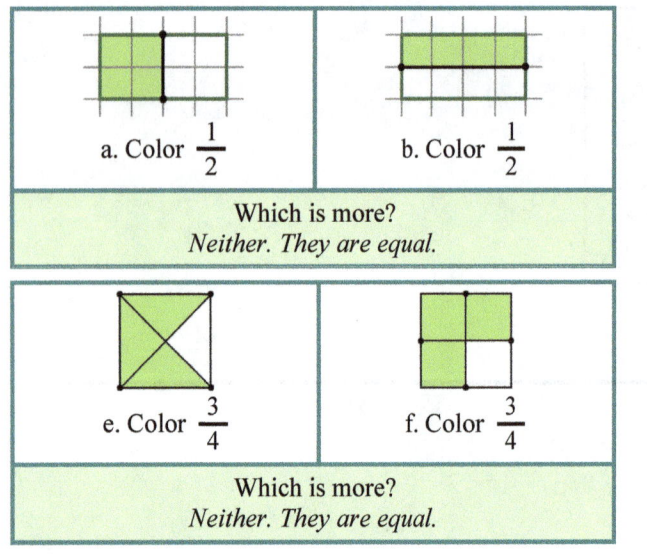 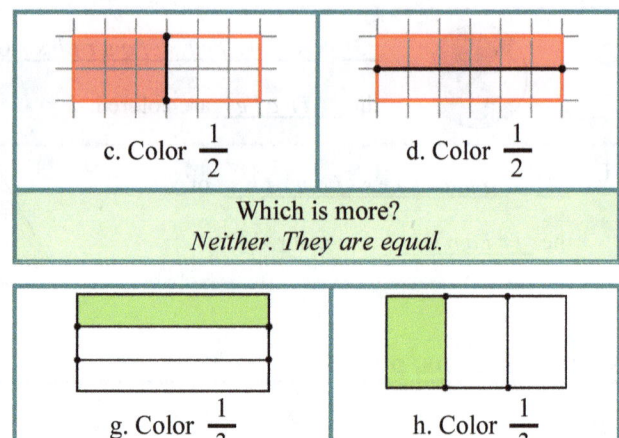

4. a. 1/2   b. 2/4   c. 2/3   d. 3/3

Understanding Fractions 1, pp. 16-19

Page 17

1. a. 1/4   b. Not equally divided   c. Not equally divided. (However, some students might notice that if you draw one more line through the square, you can divide it into eight equal parts. The shaded part is then 2/8 or 1/4 of the whole.)
   d. Not equally divided   e. 1/2

2. a. No   b. 1/6   c. No   d. No   e. 1/4

3. Answers will vary because there often are several ways to divide a shape into equal parts. For example:

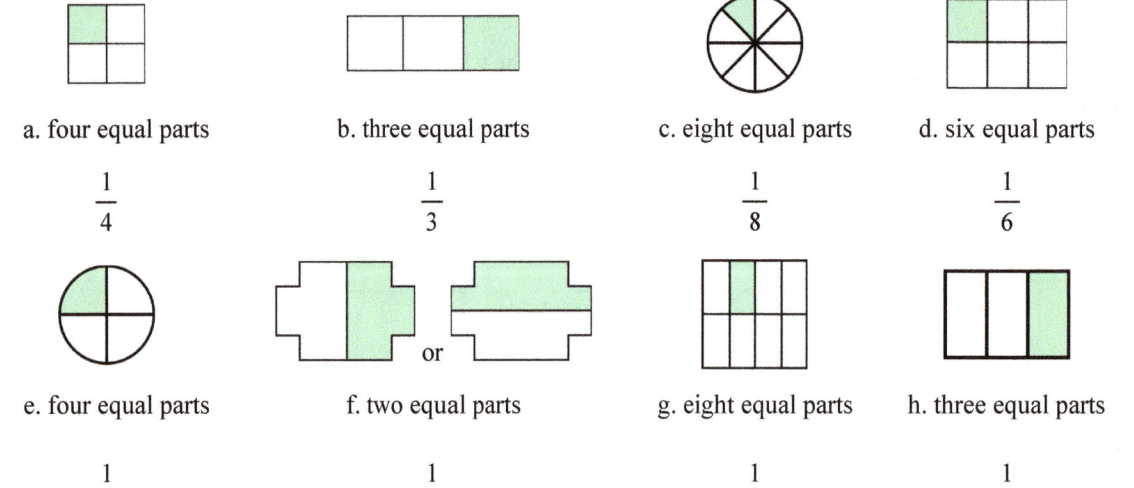

Page 18

4. Student activity. See the illustrations in the worktext.

5. The shapes *c* and *e* have six equal parts, with one part shaded.

Understanding Fractions 1, cont.

Page 18

6.

Student 1:		Student 2:	
(3×3 grid, 1 shaded) $\frac{1}{8}$	Incorrect. The answer given is the ratio of shaded parts to non-shaded parts. The correct fraction is 1/9.	(bar with 3 parts, 1 shaded) $\frac{1}{3}$	Incorrect. The parts are not all the same size. The shaded part is smaller than the non-shaded parts, so it is less than 1/3.
Student 3:		Student 4:	
(5 pentagons, 1 shaded) $\frac{1}{5}$	Correct.	(circle in 3 parts, 1 shaded) $\frac{1}{2}$	Incorrect. The answer given is the ratio of shaded parts to non-shaded parts. The correct fraction is 1/3.
Student 5:		Student 6:	
(zigzag triangles, 1 shaded) $\frac{1}{9}$	Incorrect. The triangle parts on the ends of the shape are half the size of the rest of the triangle parts, so the shaded part is less than 1/9.	(circle inside circle, inner shaded) $\frac{1}{2}$	Incorrect. The parts must be equal shape and size.

Page 19

7. a. Not equal parts   b. 1/8   c. Not equal parts

8. Answers will vary because there often are several ways to divide a shape into equal parts. For example:

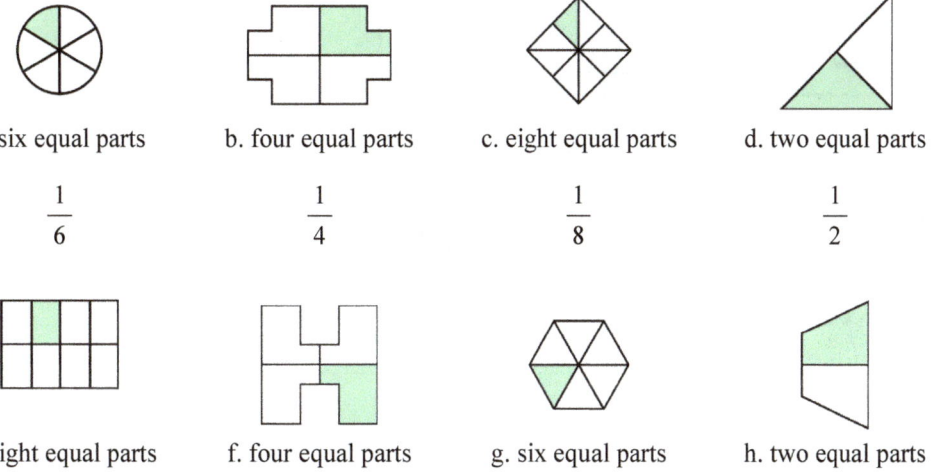

a. six equal parts   b. four equal parts   c. eight equal parts   d. two equal parts

$\frac{1}{6}$ $\frac{1}{4}$ $\frac{1}{8}$ $\frac{1}{2}$

e. eight equal parts   f. four equal parts   g. six equal parts   h. two equal parts

$\frac{1}{8}$ $\frac{1}{4}$ $\frac{1}{6}$ $\frac{1}{2}$

For 8. b., on the right is another way to divide it into four equal parts. This is based on the fact the shape itself can be divided into 20 square units.

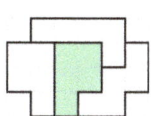

Puzzle Corner:
Yes. The large rectangle is divided into four equal rectangles, and each of those are divided equally in half. So there are eight equal parts, and the one that is shaded is 1/8 part of the shape.

Understanding Fractions 2, pp. 20-23

Page 20

1. The coloring will vary. Any parts of the whole can be colored. For example:

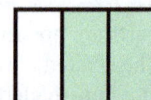 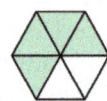 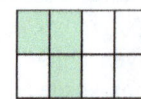 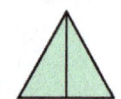 

a. 2 thirds   b. 4 sixths   c. 3 fifths   d. $\frac{3}{8}$ e. $\frac{2}{2}$ f. $\frac{9}{10}$

2. a. 1/3; one-third   b. 2/5; two-fifths   c. 3/4; three-fourths   d. 2/8; two-eighths
   e. 5/5; five-fifths   f. 3/6; three-sixths   g. 5/10; five-tenths   h. 7/8; seven-eighths

Page 21

3.
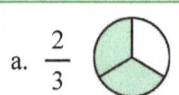   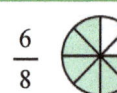

a. $\frac{2}{3}$ b. $\frac{2}{5}$ c. $\frac{3}{6}$ d. $\frac{6}{8}$

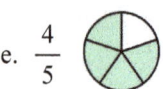  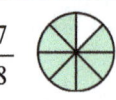

e. $\frac{4}{5}$ f. $\frac{3}{8}$ g. $\frac{3}{3}$ h. $\frac{7}{8}$

4.
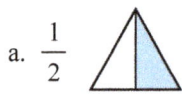 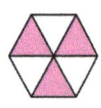 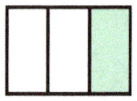 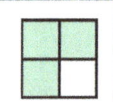

a. $\frac{1}{2}$ b. $\frac{3}{6}$ c. $\frac{1}{3}$ d. $\frac{3}{4}$

5.
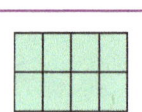 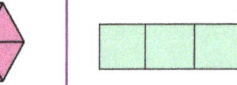

a. $1 = \frac{8}{8}$ b. $1 = \frac{6}{6}$ c. $1 = \frac{3}{3}$ d. $1 = \frac{4}{4}$ e. $1 = \frac{10}{10}$

Understanding Fractions 2, cont.

Page 22

Teaching box: You would need 8 fourths two fill two whole tables.

6.

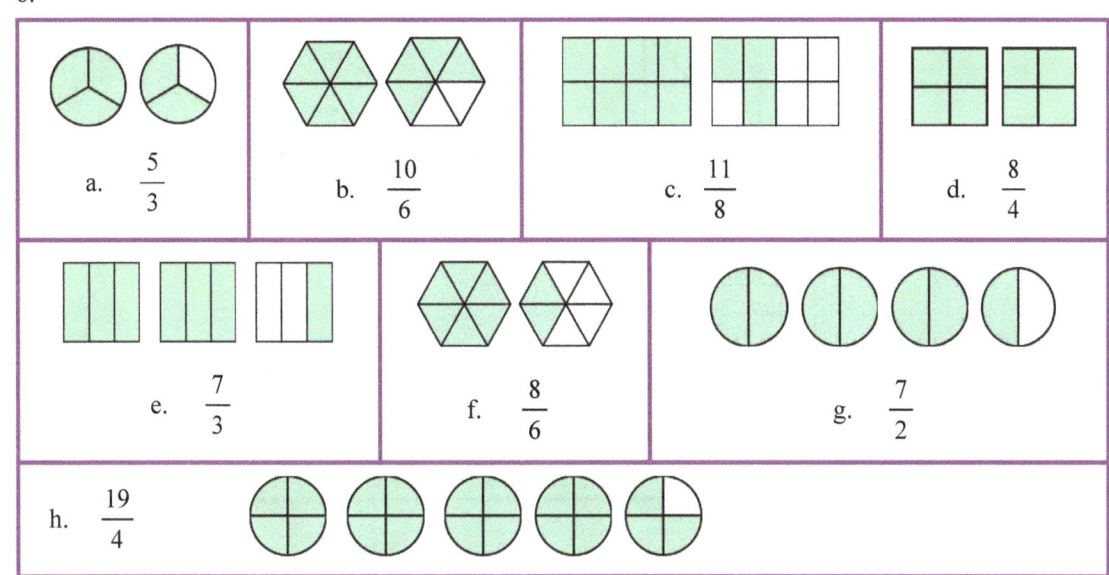

7. a. 1 half, 2 halves = 1 whole, 3 halves, 4 halves = 2 wholes,
   5 halves, 6 halves = 3 wholes, 7 halves, 8 halves = 4 wholes.

   b. 1 third, 2 thirds, 3 thirds = 1 whole, 4 thirds, 5 thirds, 6 thirds = 2 wholes,
   7 thirds, 8 thirds, 9 thirds = 3 wholes, 10 thirds, 11 thirds, 12 thirds = 4 wholes.

   c. 1 fourth, 2 fourths, 3 fourths, 4 fourths = 1 whole,
   5 fourths, 6 fourths, 7 fourths, 8 fourths = 2 wholes,
   9 fourths, 10 fourths, 11 fourths, 12 fourths = 3 wholes,
   13 fourths, 14 fourths, 15 fourths, 16 fourths = 4 wholes.

   d. 1 fifth, 2 fifths, 3 fifths, 4 fifths, 5 fifths = 1 whole,
   6 fifths, 7 fifths, 8 fifths, 9 fifths, 10 fifths = 2 wholes,
   11 fifths, 12 fifths, 13 fifths, 14 fifths, 15 fifths = 3 wholes,
   16 fifths, 17 fifths, 18 fifths, 19 fifths, 20 fifths = 4 wholes.

Page 23

8. a. 7/4; seven-fourths   b. 8/6; eight-sixths   c. 8/3; eight-thirds   d. 4/2; four-halves
   e. 5/3; five-thirds   f. 10/4; ten-fourths   g. 24/8; twenty four-eighths

9. There are several ways to draw the fractional parts together to make one whole. For example:

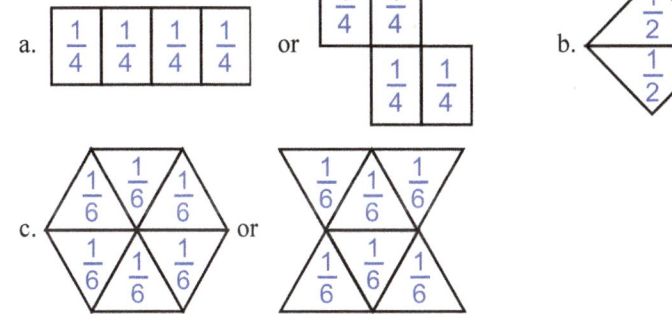

Understanding Fractions 2, cont.

Page 23

Puzzle Corner: answers will vary. For example:

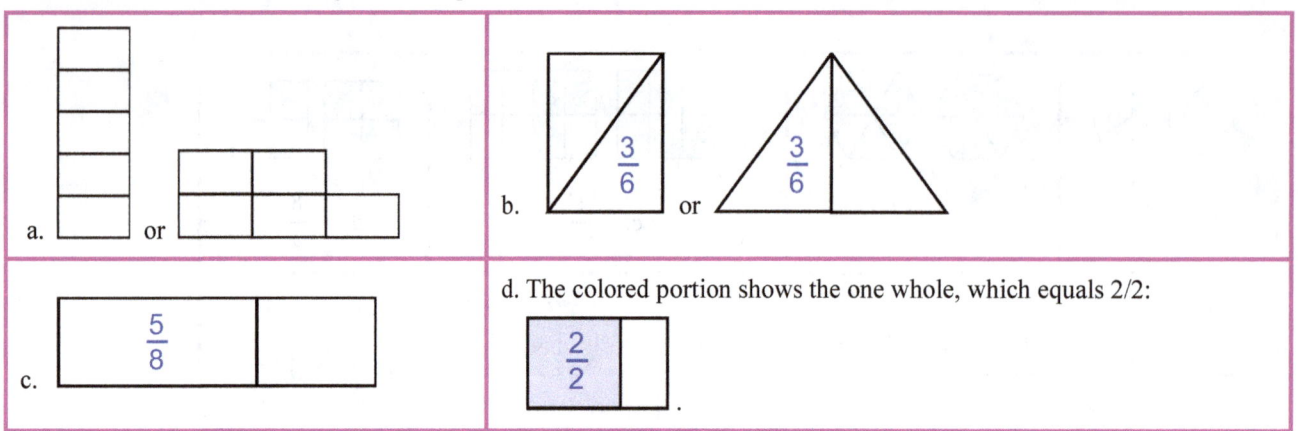

Fractions on the Number Line 1, pp. 24-26

Page 24

1. a. [number line from 0 to 1 marked 0/4, 1/4, 2/4, 3/4, 4/4] b. [number line from 0 to 1 marked 0/7, 1/7, 2/7, 3/7, 4/7, 5/7, 6/7, 7/7]

2. a. 3/8   b. 2/5
   c. 2/6   d. 5/8

Page 25

3. a. The top number line is divided into sixths.
   b. 2/6 [number line]

4. a. 2/4 [number line]   b. 3/5 [number line]
   c. 4/6 [number line]   d. 1/8 [number line]

5. a. 1/3   b. 2/9
   c. 6/10 d. 4/6

6. a. The dot marks 3/8.   b. He most likely counted the zero mark as the first fractional part.

Fractions on a Number Line, cont.

Page 26

7. a. 1/4   b. 1/3   c. 3/4   d. 1/6   e. 5/6   f. 3/8   g. 4/5   h. 5/8

8. $\frac{5}{5}$ is the biggest fraction. $\frac{5}{10}$ is the smallest fraction. See the number lines on the right.

9. The correct way to show a fraction on a number line is to mark the fraction on the mark, as it is on the top line. This is because in a number line illustration, the point for a number shows the distance from zero to that point.

If we had a rectangle (not a number line) that was divided into three parts, then labeling the section of one-third below it (like in the bottom illustration) would be correct.

Fractions on a Number Line 2, pp. 27-30

Page 27

1. a. Points at $\frac{3}{6}$, $\frac{7}{6}$, $\frac{11}{6}$, $\frac{13}{6}$, $\frac{18}{6}$.

   18/6 is the whole number 3.

   b. Points at $\frac{6}{5}$, $\frac{9}{5}$, $\frac{11}{5}$, $\frac{13}{5}$, $\frac{15}{5}$.

   15/5 is the whole number 3.

   c. Points at $\frac{5}{8}$, $\frac{12}{8}$, $\frac{16}{8}$, $\frac{17}{8}$, $\frac{21}{8}$.

   16/8 is the whole number 2.

   d. 2 = 12/6   e. 3 = 24/8

Fractions on a Number Line 2, cont.

Page 28

2.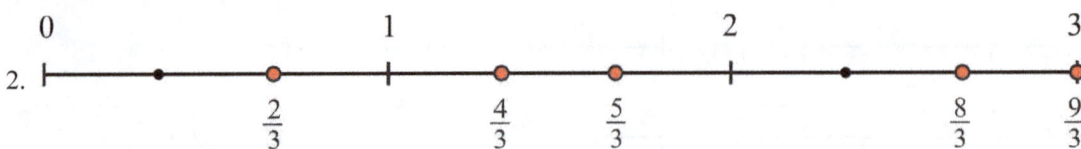

3. 4 = 12/3

4. a. 1 = 6/6   b. 2 = 16/8   c. 3 = 12/4
   d. 2 = 12/6   e. 3 = 15/5   f. 3 = 9/3   g. 2 = 16/8   h. 3 = 15/5

5. The number of fractional parts in one whole is the denominator (bottom number), and you can multiply the number of wholes by the denominator to find the top number (numerator).
   8/8 is one whole, in eighths, so 5 wholes would be (5 × 8)/8 = 40/8.
   3/3 is one whole, in thirds, so 5 wholes would be (5 × 3)/3 = 15/3.

6.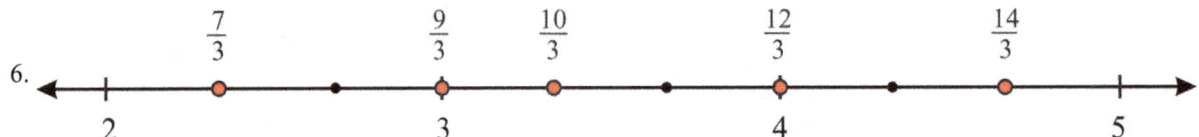

7.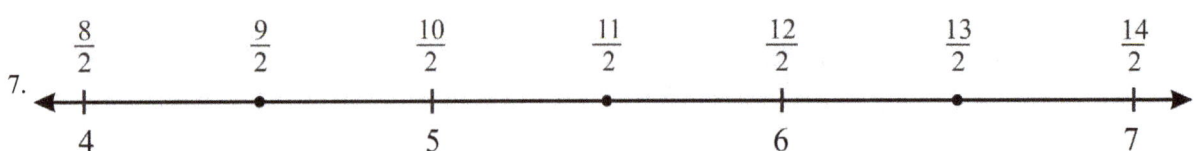

Page 29

8.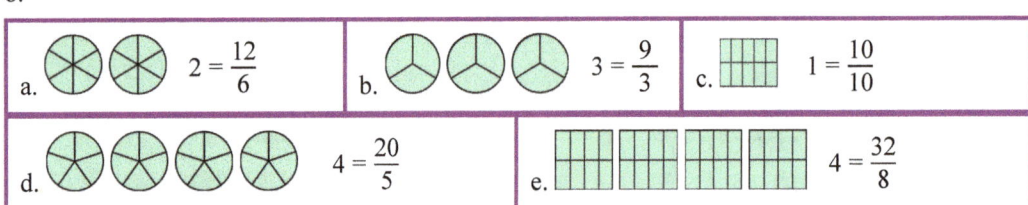

9.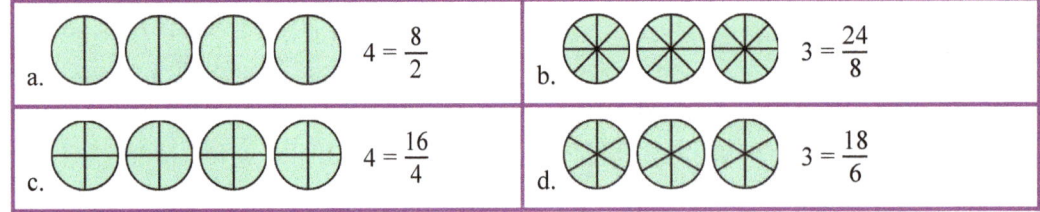

10. Any fraction where the numerator divided by the denominator equals two, will work.
    Examples: 4/2, 10/5, 20/10

11. a. 1   b. 3   c. 4   d. 10
    e. 5   f. 1   g. 4   h. 6

Page 30

12.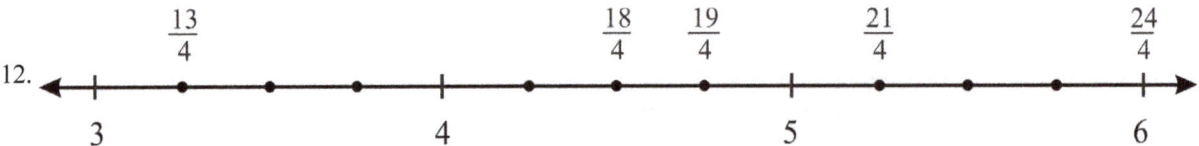

Fractions on a Number Line 2, cont.

Page 30

13.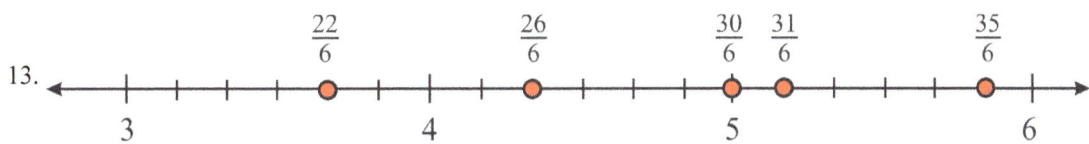

14.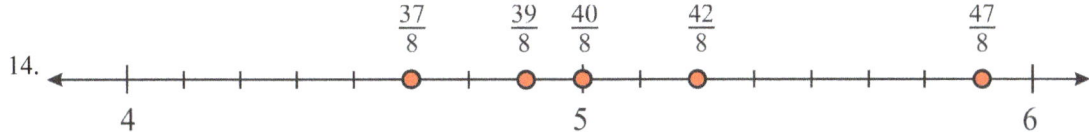

15. Neither is correct. The whole number 8 equals 64/8 and 9 equals 72/8. So, the fraction 69/8 is actually between 8 and 9.

16. 36/3 is 36 ÷ 3, which equals 12 and is the greatest number in this set.

Puzzle Corner: First, figure out where 1 is, by dividing the interval from 0 to 3-4 into three parts and thus finding the points for 1/4 and 1/2. After that, you can mark the points by fourths.

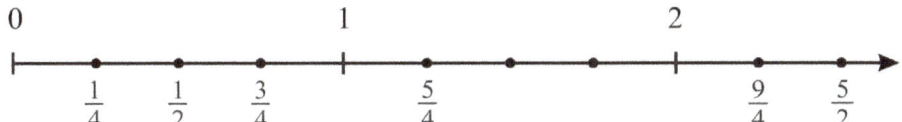

One Whole and Its Fractional Parts, pp. 31-33

Page 31

1.

a. Color 1 part.	b. Color 5 parts.	c. Color 8 parts.	d. Color 3 parts.
$\frac{1}{12}$ and $\frac{11}{12}$	$\frac{5}{10}$ and $\frac{5}{10}$	$\frac{8}{9}$ and $\frac{1}{9}$	$\frac{3}{7}$ and $\frac{4}{7}$

2.

a. $1 = \frac{9}{9}$ b. $1 = \frac{3}{3}$ c. $1 = \frac{12}{12}$ d. $1 = \frac{4}{4}$ e. $1 = \frac{5}{5}$

3. a. 1/4 of the pie is left.   b. 5/6 of the pizza is left.   c. 1/5 of the bar.

Page 32

4. a. 2/5   b. 4/5   c. 1 4/5   d. 2 3/5   e. 3 1/5   f. 4 2/5   g. 5 3/5

5. a. 2/6   b. 5/6   c. 1 1/6   d. 1 4/6   e. 2 3/6   f. 2 5/6

6. a. 1/3   b. 1 1/3   c. 1 2/3   d. 2 2/3

One Whole and Its Fractional Parts, cont.

Page 32

7.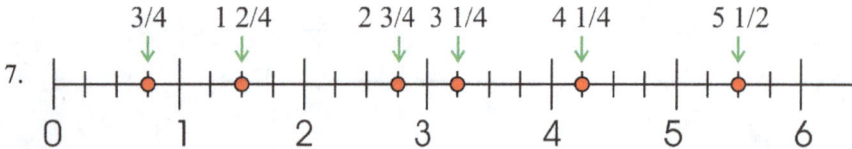

Page 33

8.

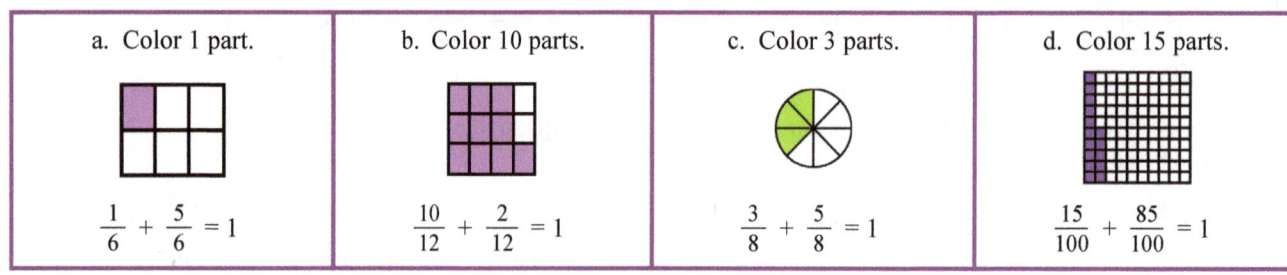

9. a. 1/4 b. 1/7 c. 7/8 d. 1/12

10. a. 2/4 or 1/2 liter
    b. 14/20 of the bread is left.

11. a. 1/10 of 90 km = 9 km. Then, 4/10 of 90 km = 36 km.
    b. First, divide $45.50 into five parts: $45.50 ÷ 5 = $9.10. Cindy pays 2/5 of the bill, or double that, which is $18.20. Sandy pays the rest, or $27.30.
    c. 7/9 is left. $2,100 is left. One-ninth of his paycheck is $300, so seven-ninths of it is 7 × $300 = $2,100.

Mixed Numbers, pp. 34-37

Page 34

1. a. 2 3/4   b. 1 1/2   c. 4 2/10
   d. 8 1/3   e. 2 4/9   f. 3 5/6

2.

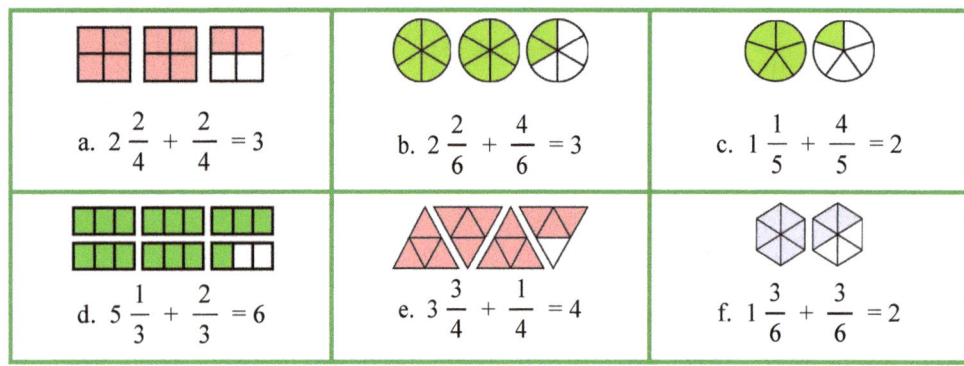

3. a. 3/4   b. 8/10   c. 5/9   d. 7/8

Mixed Numbers, cont.

Page 35

4.

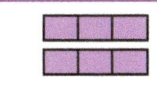	⬭⎸⬭⎸⬭⎸	⊕ ⊕ ⊕
a. 2 = 6/3	b. 3 = 6/2	c. 3 = 12/4

5.

a. $3 = \dfrac{12}{4}$	b. $1 = \dfrac{9}{9}$	c. $4 = \dfrac{4}{1}$	d. $7 = \dfrac{35}{5}$	e. $6 = \dfrac{60}{10}$
f. $7 = \dfrac{7}{1}$	g. $10 = \dfrac{60}{6}$	h. $20 = \dfrac{60}{3}$	i. $24 = \dfrac{48}{2}$	j. $50 = \dfrac{250}{5}$

6.

a. $1 = \dfrac{2}{2} = \dfrac{4}{4} = \dfrac{7}{7} = \dfrac{9}{9} = \dfrac{20}{20}$	b. $4 = \dfrac{4}{1} = \dfrac{20}{5} = \dfrac{40}{10} = \dfrac{44}{11} = \dfrac{120}{30}$

7. a. 3   b. 9   c. 30   d. 9   e. 30

Page 36

8. Answers will vary. For example:
   a. 4/5 = 1/5 + 1/5 + 2/5 or 1/5 + 3/5 or 2/5 + 2/5
   b. 5/8 = 1/8 + 1/8 + 3/8 or 2/8 + 3/8 or 2/8 + 2/8 + 1/8
   c. 2 1/3 = 2/3 + 2/3 + 3/3 or 1/3 + 4/3 + 2/3 or 1/3 + 1/3 + 5/3, etc.
   d. 1 9/12 = 6/12 + 4/12 + 11/12 or 8/12 + 5/12 + 8/12 or 4/12 + 13/12 + 4/12, etc.
   e. 2 3/6 = 4/6 + 5/6 + 6/6 or 7/6 + 4/6 + 4/6 or 3/6 + 11/6 + 1/6, etc.

Page 37

9. She can still pour 3/4 cup of water into the pitcher.

10. There is two-thirds of a pound of extra beef.

11. He had 1 7/12 of the bread left.

12. Your train of cars would be 4 1/2 inches long.

13. She needs five scoops of flour.

Puzzle corner.
a. It is not correct. You could change the total to 3 2/4 to make it correct, or change one of the addends to be 1/4 less than in the problem.
b. It is correct.

Mixed Numbers and Fractions, pp. 38-40

Page 38

1. a. 1 7/8   b. 1 2/3   c. 2 1/5
   d. 2 3/4   e. 3 4/6   f. 3 1/2

2. a. 2 3/5   b. 3 2/3   c. 5 3/4   d. 8 1/2
   e. 3 5/7   f. 6 1/9   g. 2 2/10   h. 7 2/8

Mixed Numbers and Fractions, cont.

Page 39

3. a. 13/9   b. 8/5   c. 21/8

4. a. 12/5   b. 4/3   c. 13/4   d. 9/2
   e. 21/4   f. 19/3   g. 26/3   h. 81/10

5.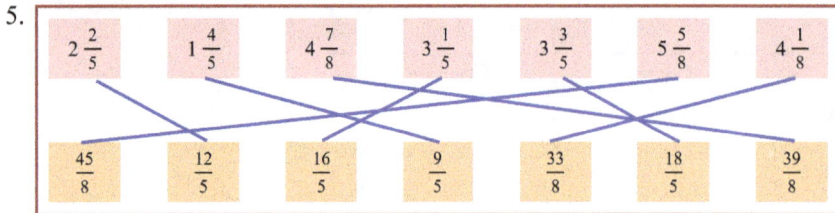

Page 40

6. a. 7 3/6   b. 22/3   c. 2 7/20   d. 23/6
   e. 5 3/4   f. 22/5   g. 4 6/7   h. 43/4

7. Answers will vary. For example:

   a. 2 5/7 = 1 + 6/7 + 6/7
      2 5/7 = 1 2/7 + 5/7 + 5/7
      2 5/7 = 4/7 + 10/7 + 5/7

   b. 2 = 5/6 + 3/6 + 4/6
      2 = 1 2/6 + 1/6 + 3/6
      2 = 4/6 + 1 1/6 + 1/6

8. The library has 2,610 ÷ 9 × 7 = 2,030 children's fiction books.
   And it has 2,610 − 2,030 = 580 children's nonfiction books.

9. a. 11/12   b. 7/9   c. 11/15   d. 13/20

10. a. 16/2   b. 70/7   c. 66/11   d. 80/4   e. 96/8

Comparing Fractions 1, pp. 41-42

Page 41

1. a. $1 = \frac{3}{3}$ "One whole is 3 thirds."   b. $1 = \frac{2}{2}$ "One whole is 2 halves."
   c. $1 = \frac{4}{4}$ "One whole is 4 fourths."   d. $1 = \frac{3}{3}$ "One whole is 3 thirds."

2.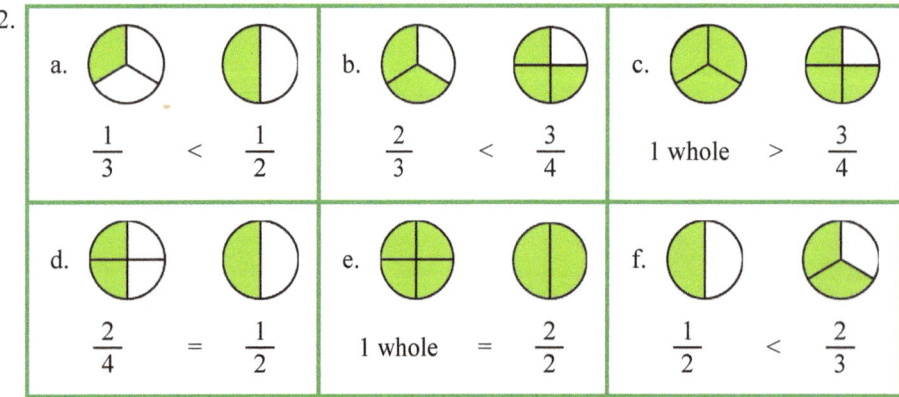

Comparing Fractions 1, cont.

Page 41

3.

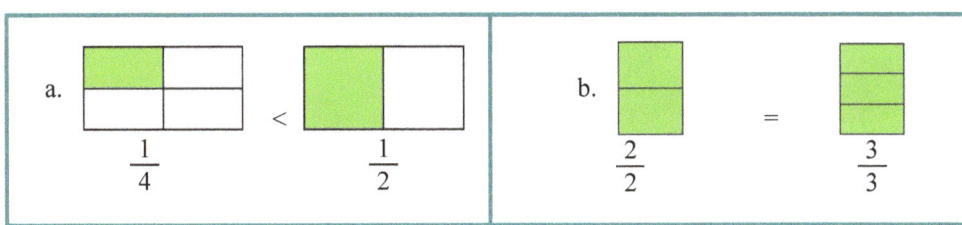

Page 42

4.

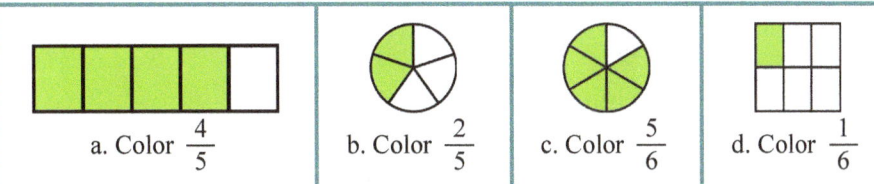

5.

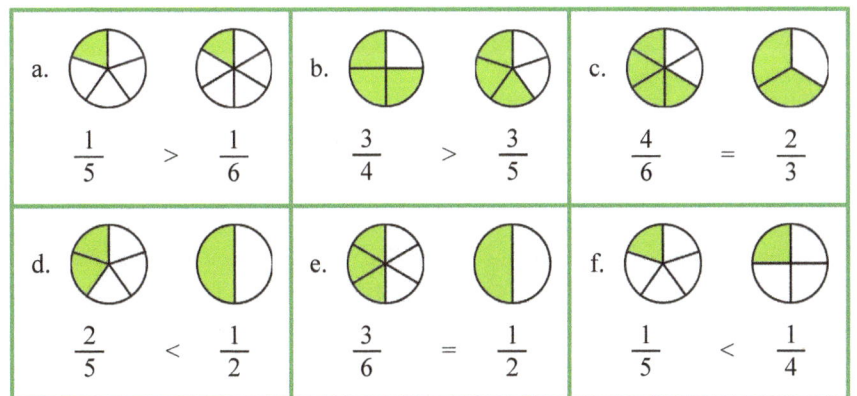

6.

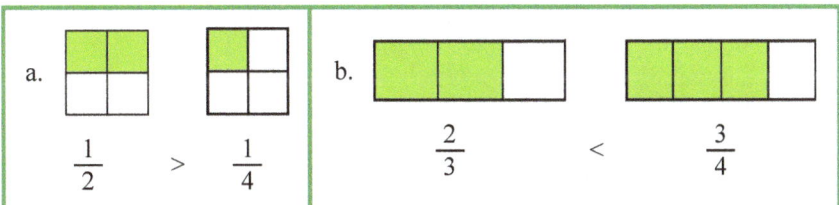

Comparing Fractions 2, pp. 43-45

Page 43

1.

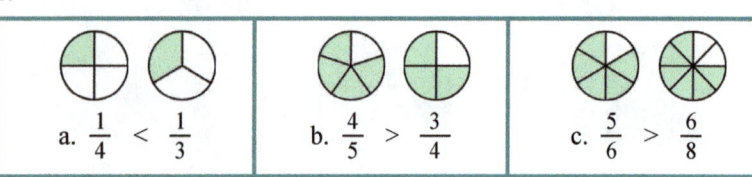

2.

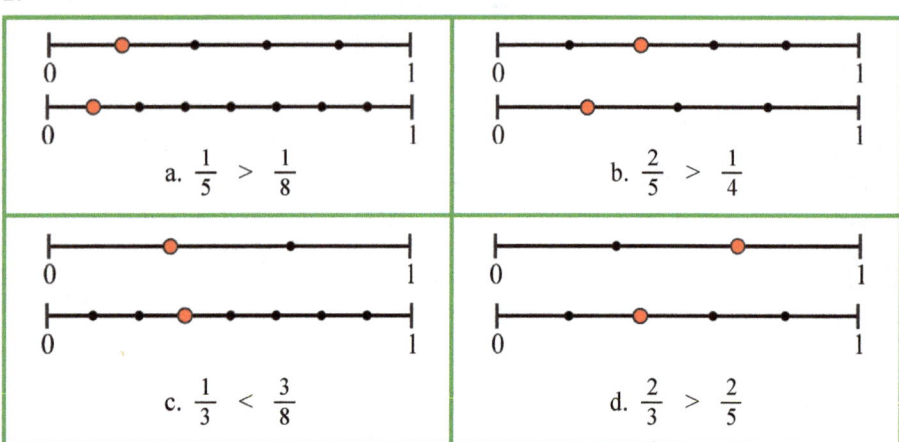

Page 44

3. a. (cross out)   b. 7/8 < 9/10
   c. 6/8 < 5/6   d. 3/8 > 3/10

4. Hazel is correct. While both fractions are equal to one whole, in this case the wholes are not the same size.

Page 45

5. a. 2/3 > 1/3   b. 1/5 < 4/5   c. 1/6 < 3/6
   d. (cross out)   e. (cross out)   f. 7/10 > 4/10

6. You can just compare how many parts there are. For example, 5/8 has more eighths than 3/8, so it is the bigger fraction. In other words, just compare the numerators. 7/6 has more sixths than 1/6, so it is bigger.

7. a. $\frac{3}{4} < \frac{5}{4}$ b. $\frac{3}{6} > \frac{1}{6}$ c. $\frac{9}{8} > \frac{8}{8}$ d. $\frac{5}{5} > \frac{2}{5}$

   e. $\frac{13}{10} > \frac{3}{10}$ f. $\frac{3}{3} < \frac{5}{3}$ g. $\frac{3}{6} < \frac{6}{6}$ h. $\frac{5}{2} > \frac{2}{2}$

8. No. The bigger pitcher has more. 1/4 of something that is larger is more than 1/4 of something that is smaller.

9. It is hard to tell who got more pie to eat. You cannot really tell. Notice that the wholes are not the same size.

Comparing Fractions 3, pp. 46-48

Page 46

1.

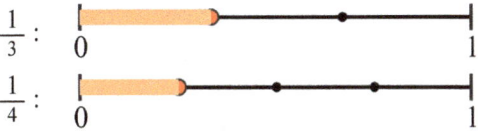

a. $\frac{1}{3} < \frac{1}{2}$	b. $\frac{1}{2} > \frac{1}{5}$
c. $\frac{1}{5} < \frac{1}{4}$	d. $\frac{1}{6} < \frac{1}{5}$
e. $\frac{1}{6} > \frac{1}{8}$	f. $\frac{1}{2} > \frac{1}{8}$

2. a. One eighth (1/8) is the bigger fraction. If you divide a whole into 8 pieces, each piece is bigger than if you divide a whole into 9 pieces.
   b. For unit fractions, the larger the denominator, the smaller the fraction.

3. Since 1/3 is further from zero than 1/4, it is the bigger fraction.

4. The number lines are not of the same length so are representing the fractions incorrectly. When we make the two number lines to have the same length from 0 to 1, we see that 1/5 < 1/4.

You can also use pie pictures: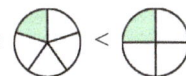

Page 47

5. a.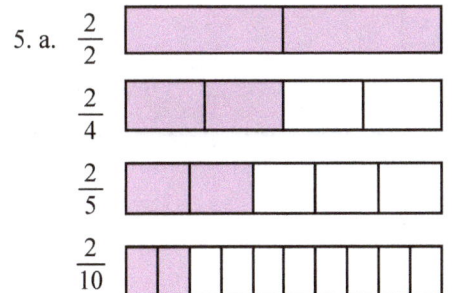

   b. The fractions all have the same numerator (number of parts). They each have a different denominator (kind of parts).

   c. The size of each fraction, compared to the others, is based on the size of its parts. When a whole is divided into more parts, the parts are smaller.

6.

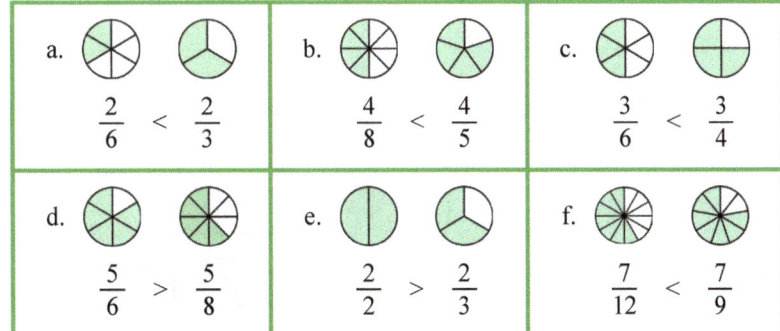

a. $\frac{2}{6} < \frac{2}{3}$	b. $\frac{4}{8} < \frac{4}{5}$	c. $\frac{3}{6} < \frac{3}{4}$
d. $\frac{5}{6} > \frac{5}{8}$	e. $\frac{2}{2} > \frac{2}{3}$	f. $\frac{7}{12} < \frac{7}{9}$

7. You can just check what kind of pieces the two fractions have, and choose the fraction that has bigger pieces. For example, 5/8 has eighths, and eighths are bigger pieces than ninths, so 5/8 is more than 5/9. Eighths are bigger pieces than tenths, so 6/8 is bigger than 6/10. In other words, just compare the denominators, and the fraction with the smaller denominator is the greater fraction.

Comparing Fractions 3, cont.

Page 48

8. a. > b. < c. > d. > e. = f. < g. = h. =

9. a. > b. > c. > d. = e. < f. = g. < h. <

10. Answers will vary. Check the student's answer. For example: Think of the fractions as so many pieces. Perhaps draw a picture. In this case, we can see that 5 thirds is more than 1, and 7 eighths is less than 1, so 5/3 is greater than 7/8.

11. $\frac{1}{9} < \frac{1}{6} < \frac{1}{5} < \frac{1}{3}$

Puzzle Corner

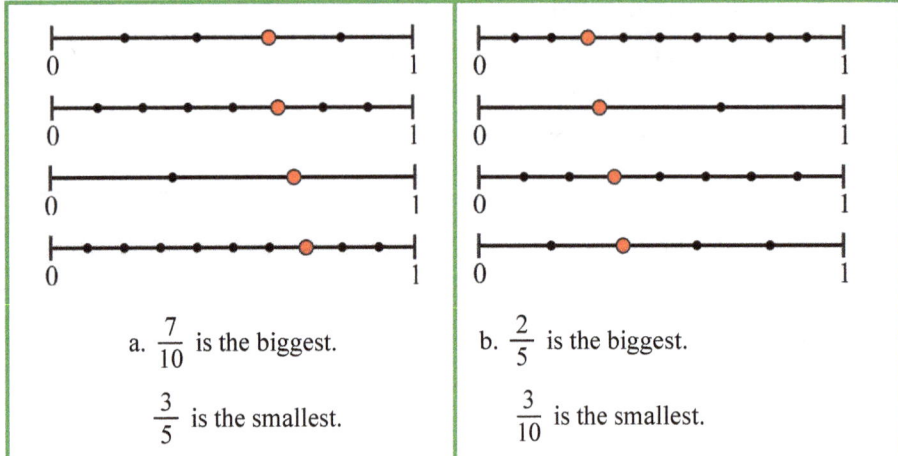

a. $\frac{7}{10}$ is the biggest.

$\frac{3}{5}$ is the smallest.

b. $\frac{2}{5}$ is the biggest.

$\frac{3}{10}$ is the smallest.

Comparing Fractions 4, pp. 49-50

Page 49

1. a. (cross out)   b. 2/9 < 2/6   c. (cross out)
   d. 3/10 > 1/4   e. (cross out)   f. 2/7 > 2/9

2. A piece from the red ribbon. That is because 1/4 > 1/5.

3. a. $\frac{4}{3} > \frac{3}{3}$ b. $\frac{6}{7} > \frac{6}{9}$ c. $\frac{9}{10} > \frac{7}{10}$ d. $\frac{9}{12} < \frac{9}{5}$

   e. $\frac{1}{6} < \frac{1}{4}$ f. $\frac{1}{12} < \frac{10}{10}$ g. $\frac{3}{8} < \frac{3}{6}$ h. $\frac{1}{2} < \frac{8}{8}$

4. The bar model for 9/10 was shorter than the one for 7/8, so it wasn't an accurate comparison.

   Now we have two wholes that are the same size. Now we can compare, and see that 9/10 is slightly greater than 7/8. We write: 9/10 > 7/8

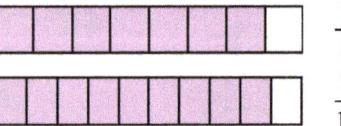

Comparing Fractions 4, cont.

Page 50

5.

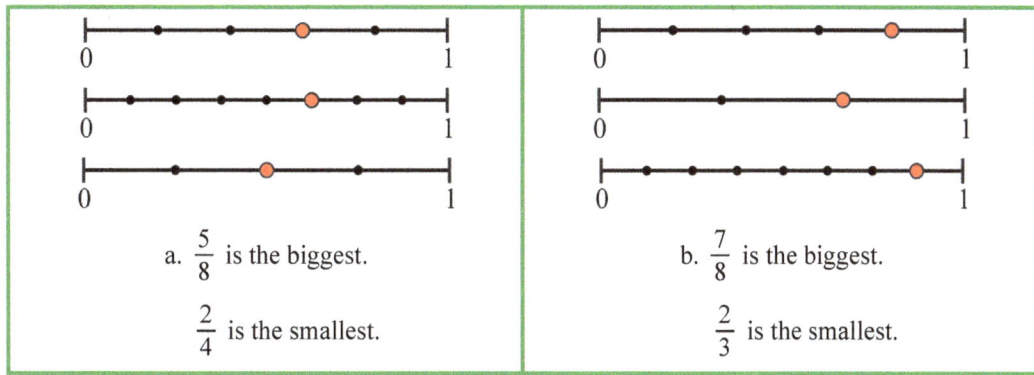

a. $\frac{5}{8}$ is the biggest.

$\frac{2}{4}$ is the smallest.

b. $\frac{7}{8}$ is the biggest.

$\frac{2}{3}$ is the smallest.

6. One-twelfth of the bigger bar is more to eat than 2/12 of the smaller. However, that does not prove that 1/12 > 2/12 because we were not using wholes of the same size.

7. No, it will not be fair. One-third of the large pizza is a bigger piece to eat than one-third of the small pizza.

Puzzle Corner: $\frac{5}{3} < \frac{8}{4} < \frac{13}{5} < \frac{18}{6} < \frac{36}{10}$

Comparing Fractions 5, pp. 51-54

Page 51

1. a. < b. > c. < d. >

2. a. > b. > c. > d. <

3. a. < b. > c. > d. > e. > f. < g. = h. <

4. a. $\frac{3}{8}, \frac{3}{6}, \frac{6}{8}$ b. $\frac{2}{5}, \frac{5}{6}, \frac{6}{5}$ c. $\frac{1}{7}, \frac{1}{4}, \frac{5}{8}$

Page 52

5. a. > b. < c. < d. <

6. a. < b. < c. < d. < e. > f. < g. > h. > i. > j. >

7.

a.	$\frac{1}{5}$ $\frac{3}{10}$ ↓ ↓ $\frac{2}{10} < \frac{3}{10}$	b.	$\frac{3}{4}$ $\frac{5}{8}$ ↓ ↓ $\frac{6}{8} > \frac{5}{8}$	c.	$\frac{5}{12}$ $\frac{1}{3}$ ↓ ↓ $\frac{5}{12} > \frac{4}{12}$	d.	$\frac{11}{12}$ $\frac{5}{6}$ ↓ ↓ $\frac{11}{12} > \frac{10}{12}$
e.	$\frac{3}{4}$ $\frac{9}{12}$ ↓ ↓ $\frac{9}{12} = \frac{9}{12}$	f.	$\frac{5}{9}$ $\frac{2}{3}$ ↓ ↓ $\frac{5}{9} < \frac{6}{9}$	g.	$\frac{1}{3}$ $\frac{2}{9}$ ↓ ↓ $\frac{3}{9} > \frac{2}{9}$	h.	$\frac{3}{12}$ $\frac{1}{3}$ ↓ ↓ $\frac{3}{12} < \frac{4}{12}$

Comparing Fractions 5, cont.

Page 53

8. a. cannot compare   b. 3/9 = 2/6   c. 7/10 > 5/8   d. cannot compare   e. cannot compare   f. cannot compare

9. $\dfrac{1}{3}, \dfrac{3}{8}, \dfrac{2}{5}, \dfrac{5}{8}, \dfrac{2}{3}$

10. Answers will vary. The student can use number lines, bars, circles, or other shapes. For example:

11.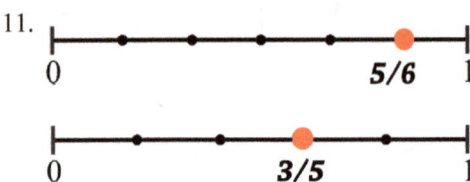

Page 54

12. Angie ate more pizza. She ate 1/8 of the pizza more than Joe. That is because Joe ate 1/4 = 2/8 of the pizza.

13. Chloe does. She pays 3/10, which is 30/100, of her paycheck in taxes.

14. If it is discounted 4/10 of its price, because 4/10 = 40/100.

15. a. The wholes are not the same size.

   b. The student can use number lines, bars, circles, or other shapes. Using pie pictures, we get

16.

a. $\dfrac{3}{7}, \dfrac{3}{5}, 1\dfrac{1}{7}$	b. $\dfrac{3}{8}, \dfrac{3}{6}, 1\dfrac{1}{4}$	c. $\dfrac{4}{9}, \dfrac{2}{3}, \dfrac{6}{5}$

Puzzle Corner. Dad ate more. Eating 2/3 of the smaller pizza, which is 1/2 the size of the larger pizza, is equal to eating 1/3 of the larger pizza. Dad ate 3/8 of the larger pizza. Now, 3/8 > 1/3 (see exercise #9), so Dad ate more pizza.

Equivalent Fractions 1, pp. 55-56

Page 55

1. a. $\dfrac{4}{6} = \dfrac{2}{3}$ b. $\dfrac{2}{3} = \dfrac{6}{9}$ c. $\dfrac{3}{6} = \dfrac{1}{2}$

   d. $\dfrac{1}{3} = \dfrac{3}{9}$ e. $\dfrac{6}{8} = \dfrac{3}{4}$

2.

a. $\dfrac{1}{4} = \dfrac{2}{8}$	b. $\dfrac{2}{4} = \dfrac{3}{6}$	c. $\dfrac{6}{8} = \dfrac{9}{12}$	d. $\dfrac{2}{3} = \dfrac{8}{12}$

Equivalent Fractions 1, cont.

Page 56

3. Illustrations may vary. For example:

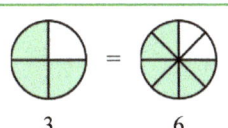

a. $\dfrac{3}{4} = \dfrac{6}{8}$ b. $\dfrac{1}{3} = \dfrac{2}{6}$

4. Fractions and illustrations may vary. Students may use pie models, rectangular models, number lines, squares, or other shapes to show this. The denominator should be twice as large as the numerator. Examples:

2/4 = 4/8 = 6/12

5. *a, b,* and *c* all show fractions equivalent to 3/4.

6. Yes. Both 3/3 and 4/4 are equal to 1, so they are equivalent fractions.

7.

a. $\dfrac{3}{4} = \dfrac{6}{8}$		b. $\dfrac{3}{5} = \dfrac{6}{10}$		c. $\dfrac{1}{3} = \dfrac{2}{6}$	
d. $\dfrac{1}{4} = \dfrac{2}{8}$		e. $\dfrac{1}{3} = \dfrac{3}{9}$		f. $\dfrac{3}{4} = \dfrac{9}{12}$	

Equivalent Fractions 2, pp. 57-58

Page 57

1. [Number lines from 0 to 2, divided into 2, 4, 8, and 16 equal parts respectively]

2. Fractions may vary. The numerator divided by the denominator should equal two. Students may use pie models, rectangular models, number lines, squares, or other shapes to show this. If the student has difficulty, point out that the number lines from question #1 can help. Examples: 2/1, 4/2, 8/4, 6/3, 10/5, 14/7, 16/8, etc.

3. a. 1/1 = 2/2 = 4/4 = 8/8   b. 1/4 = 2/8   c. 3/2 = 6/4 = 12/8

Equivalent Fractions 2, cont.

Page 58

4. Fractions may vary for *b*.

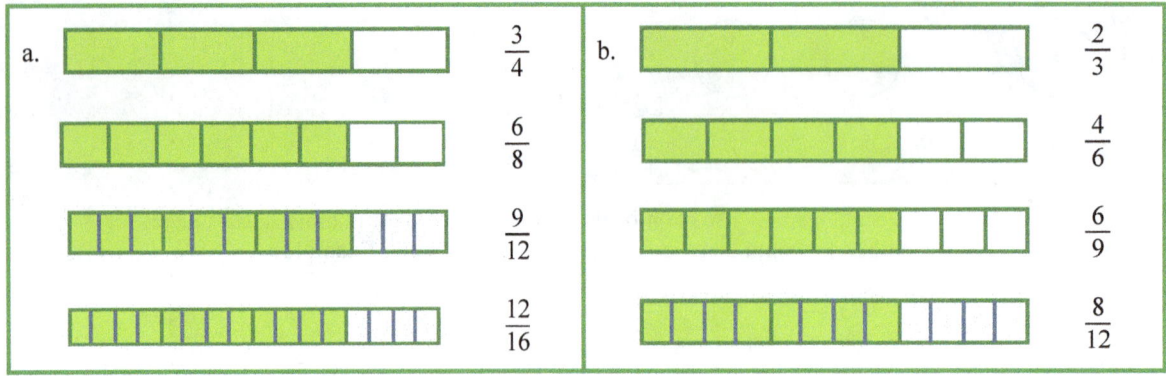

5. a. 6/2, 9/3, 3/1
   b. They can be determined to all be equal to 3, because each one, the numerator divided by the denominator equals 3.

6. Amanda ate half of her pizza. For Joe to eat half of his pizza, he would eat half of the 8 pieces. He ate 4 pieces.

7. The matching fractions are:
   1/4: Top row: 2nd, 4th, bottom row: 5th.
   1/3: Top row: 3rd, bottom row: 3rd.
   3/4: Top row: 5th, bottom row: 4th.
   2/3: Bottom row: 1st, 2nd.

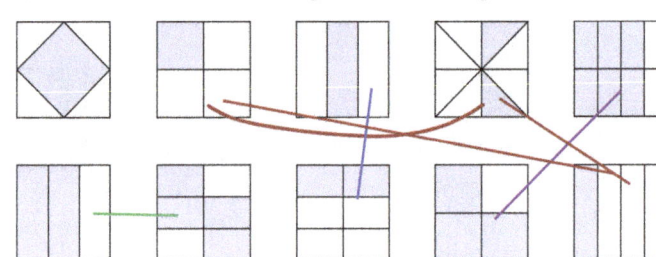

Equivalent Fractions 3, pp. 59-60

Page 59

1. 1/3 is equivalent to it.

2. The way the student corrects the false equations will vary.

   a. Not correct. Corrected version: 3/3 = 1 or 9/3 = 3 or 3/1 = 3.
   b. Not correct. Corrected version: 3/1 = 3 or 1/1 = 1 or 3/3 = 1.
   c. Correct.
   d. Correct.
   e. Not correct. Corrected version: 3 = 9/3 or 3 = 12/4 or 4 = 12/3.

3. Illustrations will vary; check the student's work. For example:

4. One easy way to show Liam he is wrong is by drawing pictures of the two fractions:

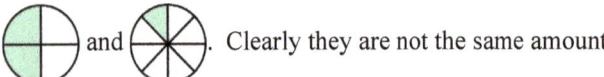 Clearly they are not the same amount.

An equivalent fraction is not formed by multiplying the numerator and the denominator. However, perhaps Liam is thinking of the equivalence of 1/4 and 2/8, and how, if you multiply both 1 and 4 by two, you get 2 and 8. That is a true process: you *can* multiply both the numerator and the denominator by some same number and that way get an equivalent fraction.

A fraction is the same as a division, so 2/4 can't be calculated as 2 × 4. 2/4 is the same as 2 divided by 4, and 1/8 is 1 divided by 8, which are not equivalent.

Equivalent Fractions 3, cont.

Page 59

5. a. 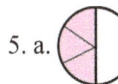 b. 1/2 = 3/6

Page 60

6.

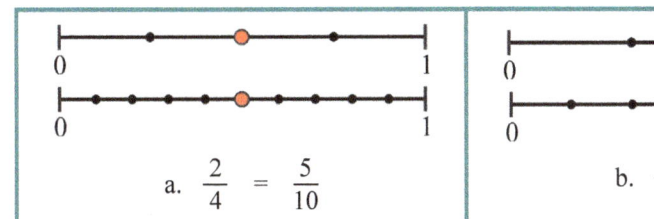

7. a. 8/2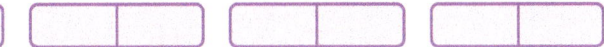

b. Answers may vary. If the bars are cut into thirds, the fraction will be 12/3. Or, like in the image below, if they are cut into fourths, the fraction will be 16/4.

  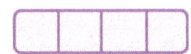

8. 2/4 = 5/10 = 4/8; 2/6 = 1/3; 4/2 = 2/1 = 8/4 = 16/8; 3/4 = 6/8; 10/2 = 5/1

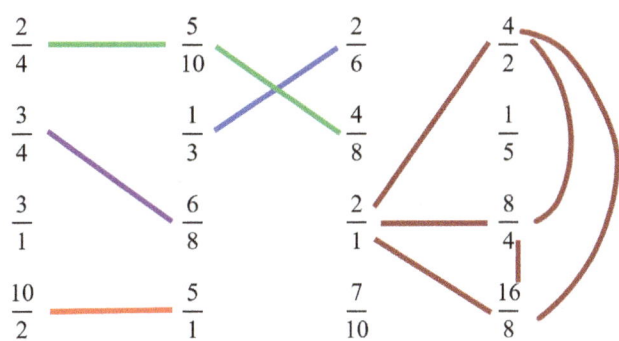

Puzzle Corner: a. 15 b. 13 c. 60 d. 25

Equivalent Fractions 4, pp. 61-65

Page 61

1.

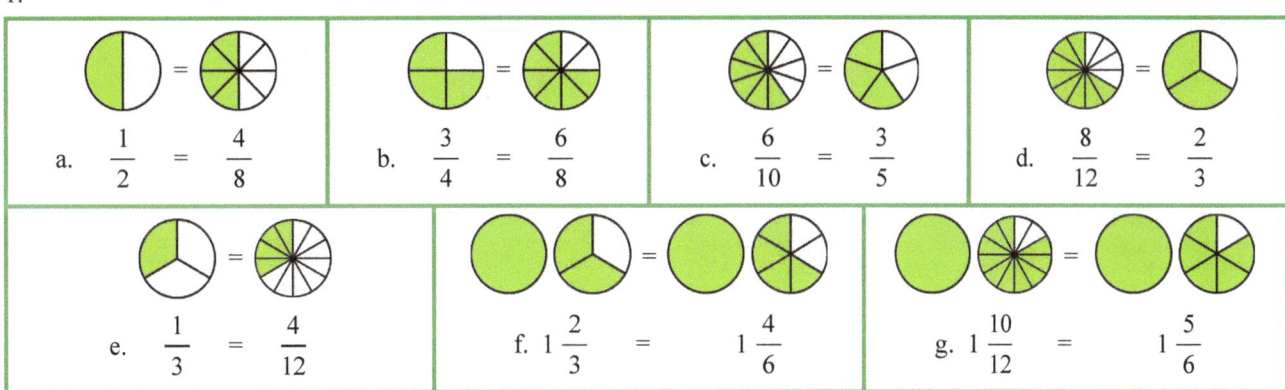

Equivalent Fractions 4, cont.

Page 61

2.

a. $\dfrac{3}{3} = \dfrac{6}{6}$	b. $\dfrac{4}{3} = \dfrac{8}{6}$	c. $\dfrac{7}{3} = \dfrac{14}{6}$
d. $2\dfrac{1}{3} = 2\dfrac{2}{6}$	e. $1\dfrac{2}{3} = 1\dfrac{4}{6}$	f. $2\dfrac{2}{3} = 2\dfrac{4}{6}$

3.

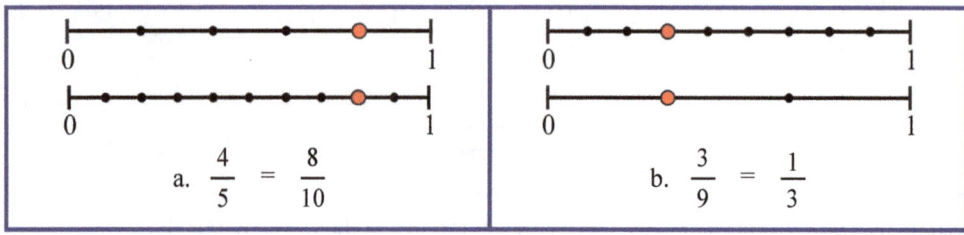

a. $\dfrac{4}{5} = \dfrac{8}{10}$ b. $\dfrac{3}{9} = \dfrac{1}{3}$

Page 62

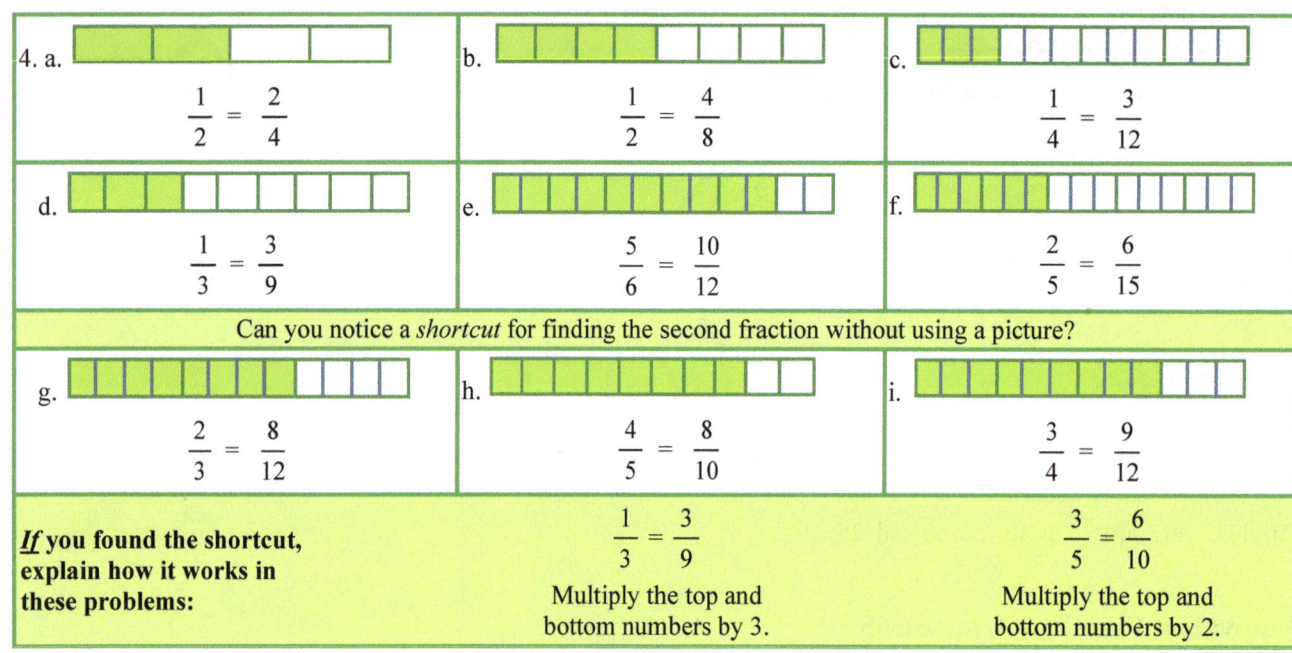

4. a. $\dfrac{1}{2} = \dfrac{2}{4}$ b. $\dfrac{1}{2} = \dfrac{4}{8}$ c. $\dfrac{1}{4} = \dfrac{3}{12}$

d. $\dfrac{1}{3} = \dfrac{3}{9}$ e. $\dfrac{5}{6} = \dfrac{10}{12}$ f. $\dfrac{2}{5} = \dfrac{6}{15}$

Can you notice a *shortcut* for finding the second fraction without using a picture?

g. $\dfrac{2}{3} = \dfrac{8}{12}$ h. $\dfrac{4}{5} = \dfrac{8}{10}$ i. $\dfrac{3}{4} = \dfrac{9}{12}$

If you found the shortcut, explain how it works in these problems:

$\dfrac{1}{3} = \dfrac{3}{9}$
Multiply the top and bottom numbers by 3.

$\dfrac{3}{5} = \dfrac{6}{10}$
Multiply the top and bottom numbers by 2.

Equivalent Fractions 4, cont.

Page 63

5.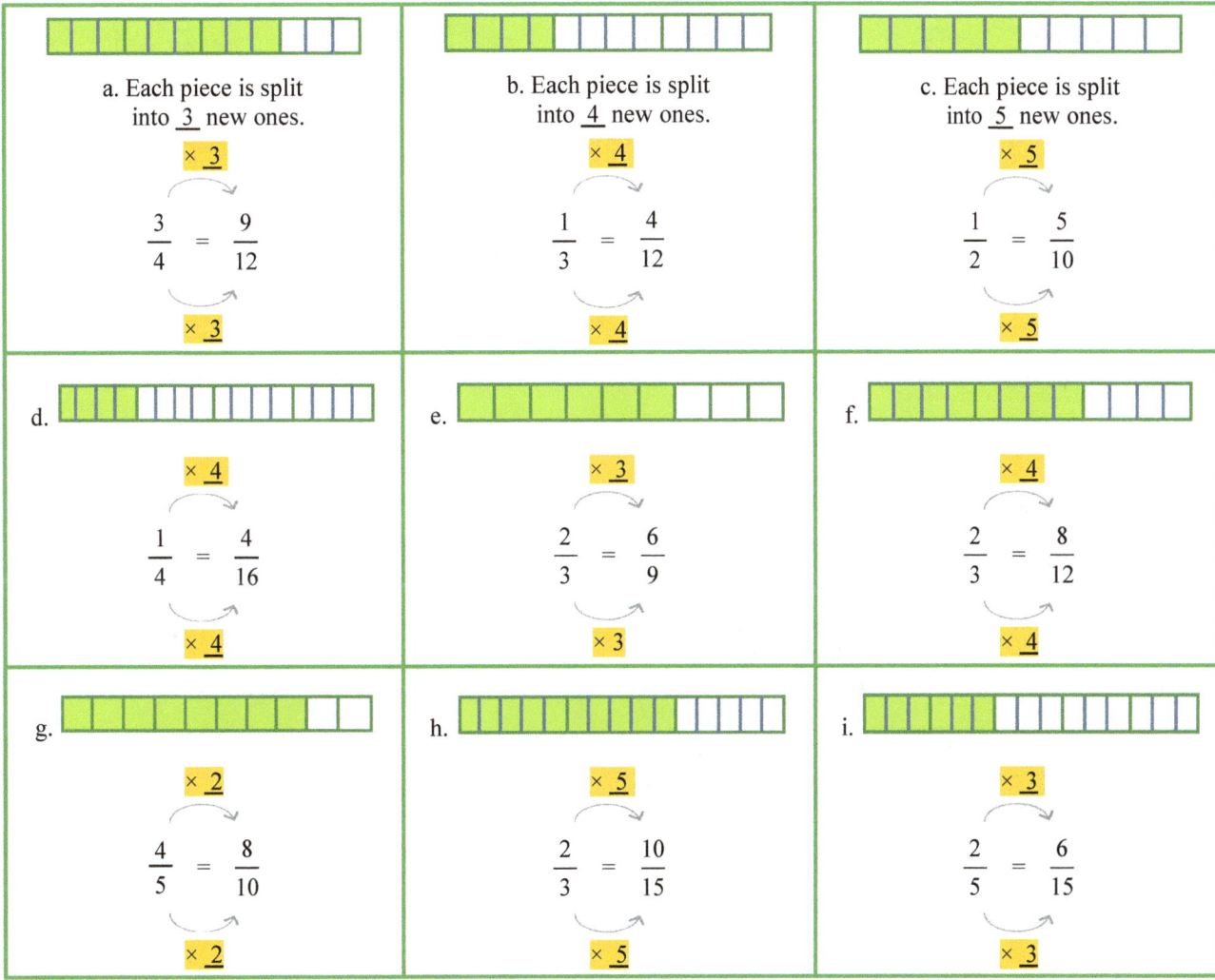

Page 64

6. a. 15/18   b. 15/20   c. 8/20   d. 90/100

7.

a. Pieces were split into 3 new ones. $\frac{1}{2} = \frac{3}{6}$	b. Pieces were split into 10 new ones. $\frac{3}{10} = \frac{30}{100}$	c. Pieces were split into 6 new ones. $\frac{2}{5} = \frac{12}{30}$	d. Pieces were split into 5 new ones. $\frac{7}{8} = \frac{35}{40}$
e. $\frac{2}{3} = \frac{4}{6}$	f. $\frac{3}{5} = \frac{9}{15}$	g. $\frac{5}{6} = \frac{10}{12}$	h. $\frac{1}{3} = \frac{3}{9}$

8.

a. $\frac{1}{10} = \frac{10}{100}$	b. $\frac{3}{10} = \frac{30}{100}$	c. $\frac{6}{10} = \frac{60}{100}$	d. $\frac{4}{10} = \frac{40}{100}$	e. $\frac{13}{10} = \frac{130}{100}$

Equivalent Fractions 4, cont.

Page 64

9.

a.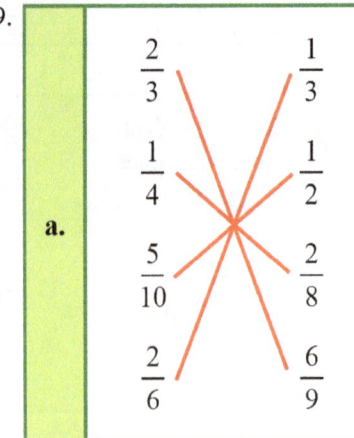
- 2/3 — 2/8
- 1/4 — 1/2
- 5/10 — 1/3 (shown as 2/6 = 1/3 via match; matches: 2/3=6/9, 1/4=2/8, 5/10=1/2, 2/6=1/3)

b.
- 1/2 = 6/12
- 3/4 = 9/12
- 1/5 = 2/10
- 4/12 = 1/3

c.
- 3/6 = 1/2
- 1/4 = 3/12
- 1/3 = 4/12
- 2/3 = 8/12

10.

a. $\frac{1}{2} = \frac{2}{4} = \frac{3}{6} = \frac{4}{8} = \frac{5}{10} = \frac{6}{12} = \frac{7}{14}$ b. $\frac{1}{3} = \frac{2}{6} = \frac{3}{9} = \frac{4}{12} = \frac{5}{15}$

Page 65

11. a. 18/100 b. 73/100 c. 75/100
 d. 99/100 e. 93/100 f. 114/100 or 1 14/100
 g. 147/100 or 1 47/100 h. 3 78/100 i. 102/100 or 1 2/100

12. Answers will vary. For example:

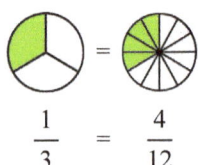

$\frac{1}{3} = \frac{4}{12}$

Puzzle corner:

a. $\frac{3}{4} + \frac{1}{2}$
 ↓ ↓
 $\frac{3}{4} + \frac{2}{4} = \frac{5}{4} = 1\frac{1}{4}$

b. $\frac{1}{5} + \frac{3}{10}$
 ↓ ↓
 $\frac{2}{10} + \frac{3}{10} = \frac{5}{10}$

c. $\frac{2}{3} + \frac{2}{9}$
 ↓ ↓
 $\frac{6}{9} + \frac{2}{9} = \frac{8}{9}$

Adding Fractions, pp. 66-67

Page 66

1.

a. $\frac{1}{6} + \frac{3}{6} = \frac{4}{6}$

b. $\frac{2}{8} + \frac{5}{8} = \frac{7}{8}$

c. $\frac{7}{8} + \frac{7}{8} = \frac{14}{8} = 1\frac{6}{8}$

d. $\frac{7}{10} + \frac{5}{10} = \frac{12}{10} = 1\frac{2}{10}$

Adding Fractions, cont.

Page 66

2.

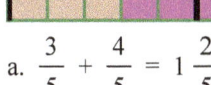

a. $\dfrac{3}{5} + \dfrac{4}{5} = 1\dfrac{2}{5}$ b. $1\dfrac{2}{5} + \dfrac{4}{5} = 2\dfrac{1}{5}$

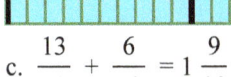

c. $\dfrac{13}{10} + \dfrac{6}{10} = 1\dfrac{9}{10}$ d. $1\dfrac{3}{8} + \dfrac{6}{8} = 2\dfrac{1}{8}$

Page 67

3. a. 2/6   b. 1   c. 7/8
   d. 1 2/5   e. 2
   f. 1 8/10   g. 1 2/4

4. 3/12 + 2/12 + 4/12 = 9/12. The children ate 9/12 of the chocolate bar. There is 3/12 left.

5. a. 2 2/5   b. 2 6/8   c. 2

6. a. 9/12   b. 3/6   c. 4/8

Adding Mixed Numbers, pp. 68-70

Page 68

1.

a. $1\dfrac{3}{5} + 2\dfrac{2}{5} = 4$ b. $1\dfrac{3}{7} + 2\dfrac{6}{7} = 4\dfrac{2}{7}$

c. $1\dfrac{3}{8} + 1\dfrac{6}{8} = 3\dfrac{1}{8}$ d. $\dfrac{8}{9} + 1\dfrac{5}{9} = 2\dfrac{4}{9}$

2. a. 5   b. 7 1/6
   c. 13 1/4   d. 10 2/8
   e. 27 2/6   f. 12 2/10

Page 69

3. These answers can be fixed in different ways. For example:

| a. In the first one, Emma has one seventh too much. The second one is correct.

$1\dfrac{5}{7} = \dfrac{2}{7} + 1\dfrac{1}{7} + \dfrac{2}{7}$

$1\dfrac{5}{7} = \dfrac{10}{7} + \dfrac{2}{7}$ | b. In the first one, Peter is lacking one third from the addition. In the second one, he has one third too much.

$2\dfrac{1}{3} = \dfrac{2}{3} + \dfrac{2}{3} + \dfrac{2}{3} + \dfrac{1}{3}$

$2\dfrac{1}{3} = \dfrac{5}{3} + \dfrac{2}{3}$ |

Adding Mixed Numbers, cont.

Page 69

4. Answers will vary. For example:
   a. 1 3/10 = 1 + 3/10 or 5/10 + 8/10 or 4/10 + 9/10 or 2/10 + 11/10
   b. 3 1/5 = 4/5 + 12/5 or 10/5 + 6/5 or 8/5 + 8/5 or 1/5 + 15/5

5. a. 1 1/2 + 1/2 + 1/2 = 2 1/2. The recipe calls for 2 1/2 cups of flour.
   b. 1 3/4 + 1 1/4 = 3. They took three hours.
   c. 1 1/2 + 3/4 = 2. He drank 2 cups of liquid.

Page 70

6. 2 1/4 + 3 1/4 + 2 1/4 + 3 1/4 = 11. Its perimeter is 11 inches.

7. 2 3/8 + 2 3/8 + 2 3/8 = 6 9/8 = 7 1/8. Its perimeter is 7 1/8 inches.

8. For the amount of sugar, 1 2/4 is also acceptable, and for the amount of flour, 2 2/4 cups is also acceptable, as students have not yet been taught how to simplify fractions.

> A birthday cake
> 8 eggs
> 1 1/2 (or 1 2/4) cups sugar
> 2 1/2 (or 2 2/4) cups flour
> 3 tsp baking powder
> 2 cups whipped cream
> sliced fruit

Puzzle corner:
a. 1 1/5   b. 1 3/4   c. 1 3/6

Subtracting Fractions and Mixed Numbers, pp. 71-74

Page 71

1. a. 8/10   b. 4/12   c. 2 2/6
   d. 1 2/9   e. 6/4 = 1 2/4   f. 2 4/8
   g. 3 4/12   h. 2/10   i. 3 4/12   j 1/8

2. a. 3/6   b. 6/10   c. 3/8   d. 3/5

Page 72

3.

| a. $3\frac{2}{10} - \frac{6}{10}$ $\downarrow \quad \downarrow$ $2\frac{12}{10} - \frac{6}{10} = 2\frac{6}{10}$ | b. $2\frac{1}{7} - \frac{5}{7}$ $\downarrow \quad \downarrow$ $1\frac{8}{7} - \frac{5}{7} = 1\frac{3}{7}$ | c. $5\frac{3}{9} - 2\frac{7}{9}$ $\downarrow \quad \downarrow$ $4\frac{12}{9} - 2\frac{7}{9} = 2\frac{5}{9}$ | d. $7\frac{2}{5} - 4\frac{4}{5}$ $\downarrow \quad \downarrow$ $6\frac{7}{5} - 4\frac{4}{5} = 2\frac{3}{5}$ |

4. a. 6 1/4   b. 5 1/9   c. 3 1/2   d. 7 2/5

5. There are 2 9/12 of the pies left.

6. a. 1 1/5   b. 1 2/5   c. 2/5

Page 73

7. a. 1 2/4   b. 2 6/8   c. 3 2/6
   d. 4/5   e. 4 3/5   f. 2 2/3

8. It is 1 inch. Half of the perimeter is 2 ½ inches, and the two sides add up to that (1 in + 1 ½ in = 2 ½ in).

Subtracting Fractions and Mixed Numbers, cont.

Page 74

9.

a. $\dfrac{6}{10} - \dfrac{15}{100}$ ↓ ↓ $\dfrac{60}{100} - \dfrac{15}{100} = \dfrac{45}{100}$	b. $\dfrac{7}{10} - \dfrac{38}{100}$ ↓ ↓ $\dfrac{70}{100} - \dfrac{38}{100} = \dfrac{32}{100}$	c. $\dfrac{54}{100} - \dfrac{2}{10}$ ↓ ↓ $\dfrac{54}{100} - \dfrac{20}{100} = \dfrac{34}{100}$

10. 6 − 2 2/3 = 3 1/3. The part left is 3 1/3 yards long. In feet, 2 2/3 yards = 8 feet, and 3 1/3 feet = 10 feet.

11. 11/12 of a pizza is left. Edward, Abigail, Jack, and John ate 13 pieces, which is 1 1/12 of a pizza. Since Mom and Dad ate 1 pizza, in total 2 1/12 of the pizzas were consumed. So, 11/12 of a pizza is left.

12. a. 6 − 2 2/3 = 3 1/3. There is 3 1/3 cups of flour left.
    b. She can make one more batch.

Puzzle corner. Half of the perimeter is 3 ¼ inches. We can write the addition: 1 ¾ + (?) = 3 ¼. Solution: (?) = 1 2/4 or 1 ½. The other side is 1 ½ inches.

Multiplying Fractions by Whole Numbers, pp. 75-77

Page 75

1.

a. $\dfrac{3}{7} = 3 \times \dfrac{1}{7}$	b. $\dfrac{6}{9} = 6 \times \dfrac{1}{9}$	c. $4 \times \dfrac{1}{5} = \dfrac{4}{5}$	d. $7 \times \dfrac{1}{10} = \dfrac{7}{10}$

2.

a. $\dfrac{8}{7} = 8 \times \dfrac{1}{7}$	b. $1 \dfrac{3}{5} = \dfrac{8}{5} = 8 \times \dfrac{1}{5}$	c. $1 \dfrac{2}{3} = \dfrac{5}{3} = 5 \times \dfrac{1}{3}$
d. $10 \times \dfrac{1}{6} = \dfrac{10}{6} = 1 \dfrac{4}{6}$	e. $7 \times \dfrac{1}{4} = \dfrac{7}{4} = 1 \dfrac{3}{4}$	f. $9 \times \dfrac{1}{3} = \dfrac{9}{3} = 3$

3. a. 10 × 1/3 = 10/3 = 3 1/3. <u>She needs to buy at least 3 1/3 lb of chicken.</u>
   b. Between 3 and 4.
   c. 10 × 1/2 = 5. <u>She needs 5 quarts of juice.</u>

Page 74

4.

a. $3 \times \dfrac{2}{4} = \dfrac{6}{4} = 1 \dfrac{2}{4}$	b. $4 \times \dfrac{2}{6} = \dfrac{8}{6} = 1 \dfrac{2}{6}$	c. $2 \times \dfrac{7}{8} = \dfrac{14}{8} = 1 \dfrac{6}{8}$

Multiplying Fractions by Whole Numbers, cont.

Page 76

5.

a.  $5 \times \frac{3}{8} = \frac{15}{8} = 1\frac{7}{8}$	b. $4 \times \frac{2}{5} = \frac{8}{5} = 1\frac{3}{5}$	
c. $5 \times \frac{7}{12} = \frac{35}{12} = 2\frac{11}{12}$	d. $5 \times \frac{6}{10} = \frac{30}{10} = 3$	
e. $9 \times \frac{5}{8} = \frac{45}{8} = 5\frac{5}{8}$		

Can you find a shortcut for these problems?
Answers will vary. For example: Multiply the top number of the fraction by the whole number.

f. $4 \times \frac{2}{3} = \frac{8}{3} = 2\frac{2}{3}$	g. $3 \times \frac{4}{10} = \frac{12}{10} = 1\frac{2}{10}$	h. $2 \times \frac{5}{6} = \frac{10}{6} = 1\frac{4}{6}$

6.

a. $\frac{8}{5} = 4 \times \frac{2}{5}$	b. $\frac{9}{4} = 3 \times \frac{3}{4}$	c. $2\frac{2}{3} = 2 \times 1\frac{1}{3}$

Page 77

7. a. 1 1/4   b. 2   c. 1 1/7   d. 1 4/10   e. 2 2/8   f. 14/100
   g. 2 1/10   h. 72/100   i. 3 3/10   j. 3 1/8   k. 2 2/3   l. 3 2/4

8. 4 × 7/8 in. = 28/8 in. = 3 4/8 in. (which is also equal to 3 1/2 in.)

9. a. 5 × 1 1/8 in. = 5 5/8 in.
   b. Double the previous result to get 10 10/8 in. = 11 2/8 in.

10. Meat: 8 × 1/4 lb = 2 lb. Pasta: 8 × 3/4 C = 24/4 C = 6 C.

Practicing with Fractions, pp. 78-79

Page 78

1.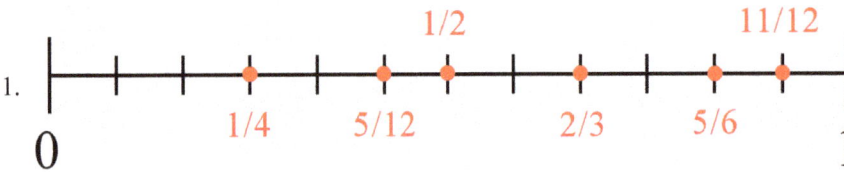

2.

a. $\frac{2}{6}, \frac{1}{2}, \frac{2}{3}$	b. $\frac{1}{8}, \frac{1}{4}, \frac{3}{8}$
c. $\frac{2}{5}, \frac{1}{2}, \frac{3}{5}$	d. $\frac{3}{8}, \frac{3}{4}, \frac{4}{5}$

3. 3 × 3/5 mi. = 9/5 mi. = 1 4/5 mi.

Practicing with Fractions, cont.

Page 78

4. Answers may vary.

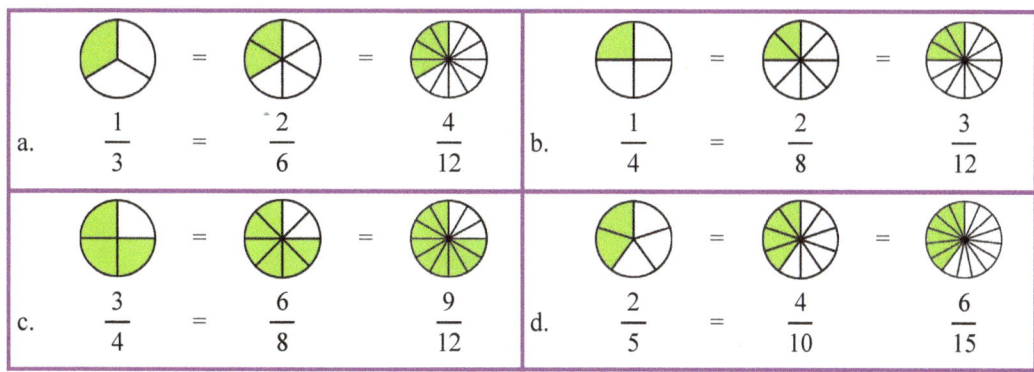

Page 79

5. a. 4/7   b. 3 1/3
   c. 2 1/4   d. 2
   e. 7   f. 2

6. a. 1   b. 1 3/8
   c. 7 1/5   d. 8/12
   e. 8/10   f. 4 2/4

7. Multiplying by one-half is actually the same as <u>dividing</u> by 2. Fill in.

a.	b.	c.
$2 \times \frac{1}{2} = 1$	$7 \times \frac{1}{2} = 3\frac{1}{2}$	$15 \times \frac{1}{2} = 7\frac{1}{2}$
$3 \times \frac{1}{2} = 1\frac{1}{2}$	$8 \times \frac{1}{2} = 4$	$20 \times \frac{1}{2} = 10$
$4 \times \frac{1}{2} = 2$	$9 \times \frac{1}{2} = 4\frac{1}{2}$	$17 \times \frac{1}{2} = 8\frac{1}{2}$
$5 \times \frac{1}{2} = 2\frac{1}{2}$	$10 \times \frac{1}{2} = 5$	$21 \times \frac{1}{2} = 10\frac{1}{2}$
$6 \times \frac{1}{2} = 3$	$11 \times \frac{1}{2} = 5\frac{1}{2}$	$32 \times \frac{1}{2} = 16$

Puzzle Corner.
a. 1/2 + 3/8 = 4/8 + 3/8 = 7/8
b. 1/3 + 1/6 = 2/6 + 1/6 = 3/6
c. 1/3 + 2/9 = 3/9 + 2/9 = 5/9

Fractions Review, pp. 80-82

Page 80

1.
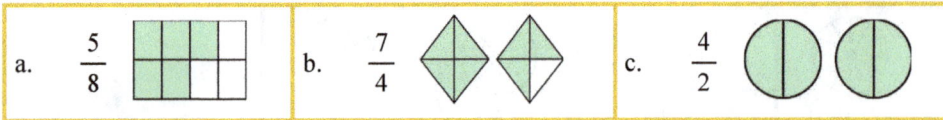

2. a. 3/6   b. 13/8   c. 6/3   d. 3/2

3. a. 1/5   b. 6/8

4. a.    b.

5.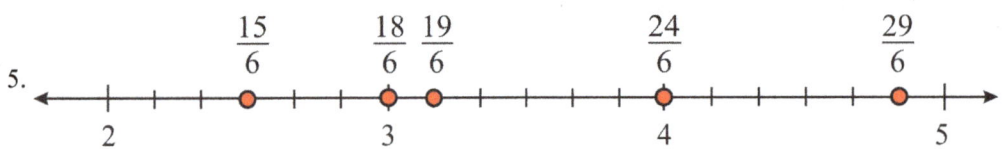

6. Since 30/6 = 5 and 36/6 = 6, 35/6 is between 5 and 6.

Page 81

7. a. 1 = 2/2   b. 2 = 20/10   c. 2 = 8/4
   d. 4 = 20/5   e. 7 = 42/6   f. 4 = 40/10   g. 6 = 12/2   h. 5 = 40/8

8.
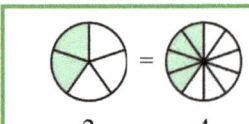

9.
$\frac{1}{3} = \frac{3}{9} = \frac{2}{6}$
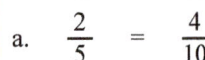

10. Answers will vary. Students may use pie models, rectangular models, number lines, squares, or other shapes to show this. For example:

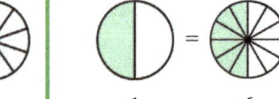

11. a. 7/10 < 3/4   b. 5/8 < 3/4   c. (cross out)

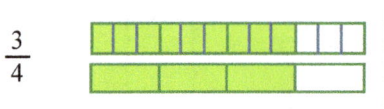

Page 82

12. a. 6/8 < 7/8   b. 1/5 > 1/8   c. 3/8 < 3/5   d. 1/2 = 2/4   e. 24/10 > 15/10

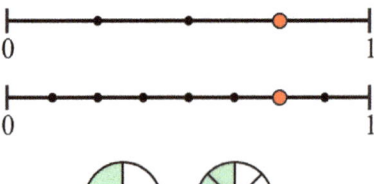

13. Looking at the numerators, we can see that both fractions have 5 pieces. Then considering the denominators, or the kind of pieces, halves are greater than sixths, so 5/2 is larger than 5/6.

14. $\frac{2}{9} < \frac{1}{3} < \frac{4}{9} < \frac{3}{6}$

Fractions Review, cont.

Page 82

15. a. The bigger can has more paint.
 b. In this case, it does not, because the wholes (paint cans) are not the same size.

Puzzle Corner: The particular squares that are shaded may vary. Check that it is the correct fractional part. For example:

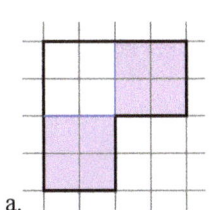
a.
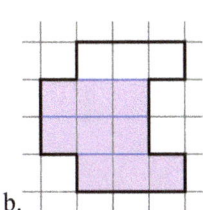
b.
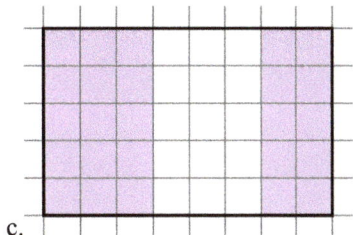
c.

Review, pp. 83-84

Page 83

1. a. 1   b. 5 1/8   c. 7   d. 2/10   e. 1 2/4   f. 6 7/12

2. a. 7/10   b. 3/5   c. 4/5   d. 5/8

3. a. 33/100   b. 53/100   c. 1 17/100

4. 1 3/4 liters

5. Answers will vary. For example: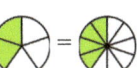

6.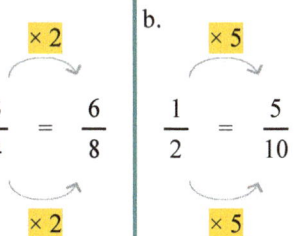

a. ×2, 3/4 = 6/8, ×2	b. ×5, 1/2 = 5/10, ×5	c. $\frac{2}{5} = \frac{4}{10}$	d. $\frac{2}{3} = \frac{6}{9}$
		e. $\frac{2}{3} = \frac{8}{12}$	f. $\frac{3}{4} = \frac{12}{16}$

Page 84

7. a. >   b. <   c. =   d. >   e. >   f. >   g. >   h. <

8. a. 9/10   b. 1 1/5   c. 1 4/10   d. 99/100   e. 2 4/8   f. 2 9/12

9.
> Mexican Coffee (4x)
>
> 6 cups strong gourmet coffee
> 3 tsp cinnamon
> 16 tsp chocolate syrup
> 1 tsp nutmeg
> 2 cup heavy cream
> 4 tbsp sugar

10. a. 40; 80   b. 8 cm; 40 cm   c. 400 kg; $40

11. Since he has 1/4 of his birthday money left, he has $5 left.

12. One-eighth of 240 pages is 30 pages. She has 5/8 of the book left to read, which is 5 × 30 = 150 pages.

Math Mammoth has a variety of resources to fit your needs. All are available as economical downloads, and most also as printed copies.

- **Math Mammoth Light Blue Series**
  A complete curriculum for grades 1-8. Each grade level includes two student worktexts (A and B), which contain all the instruction and exercises all in the same book, answer keys, tests, cumulative reviews, and a worksheet maker. International (all metric), Canadian, and South African versions are also available.
  https://www.MathMammoth.com/complete-curriculum
  https://www.MathMammoth.com/international/international
  https://www.MathMammoth.com/canada/
  https://www.MathMammoth.com/south_africa/

- **Math Mammoth Skills Review Workbooks**
  These workbooks are intended to be used alongside the Light Blue series full curriculum, and they provide additional review to the topics studied in the main curriculum, in a spiral manner.
  https://www.MathMammoth.com/skills_review_workbooks/

- **Math Mammoth Blue Series**
  Blue Series books are topical worktexts for grades 1-8, containing both instruction and exercises. They cover all elementary math topics from 1st through 8th grade and some for 8th grade. These books are not tied to grade levels, and are thus great for filling in gaps.
  https://www.MathMammoth.com/blue-series

- **Make It Real Learning**
  These activity workbooks concentrate on answering the question, "Where is math used in real life?" The series includes various workbooks for grades 3-12.
  https://www.MathMammoth.com/worksheets/mirl/

- **Review Workbooks**
  Workbooks for grades 1-8 that provide a comprehensive review of one grade level of math —for example, for review during school break or summer vacation.
  https://www.MathMammoth.com/review_workbooks/

Free gift!

- Receive over 350 free sample pages and worksheets from my books, plus other freebies:
  https://www.MathMammoth.com/free/

Lastly...

- Inspire4 is an inspirational website for the whole family I've been privileged to help with:
  https://www.inspire4.com